BIOLOGY 21
LABORATORY MANUAL VERSION 2.0
TABLE OF CONTENTS

Lab 0: Orientation

Lab 1: Cells & Tissues

Lab 2: Enzymes & Digestion

Lab 3: Skeletomuscular System

Lab 4: Nervous System & Sensory Reception

Lab 5: The Circulatory System

Lab 6: Respiration & Gas Exchange

Lab 7: Urinary & Endocrine Systems

Lab 8: The Immune Response

Lab 9: Reproduction & STDs

Lab 10: Development & Mitosis

Lab 11: Genetic Variation & Meiosis

BIOLOGY 21
LAB 0: ORIENTATION

Welcome to the Biology 21 Multimedia Lab in 448 Duncan Hall!

Write your name here (***required***): _____

PART 1: PROCEDURES

1. <u>You must have your lab manual with you to work in the lab. Sharing lab manuals is not permitted, nor are photocopies of the lab</u>. **No exceptions!!!**

2. See lab instructor for a lab card. You will need a student ID/driver's license/State ID/military ID to reserve your spot while you're in the lab. <u>You need to get a card even if you do not plan on using the computers</u>. The maximum number of people allowed in the Bio 21 lab is 33 including the lab instructor. If all the cards are taken, sign up on the waitlist. Due to College of Science regulations, you must wait in the hallway until a card becomes available.

3. Lab is self-paced. After you check in, open your lab book and get started! We encourage you to work with a fellow Bio 21 student at each of the stations and/or for the activities. We expect that the answers/responses you write in your lab manual, however, will be your own – don't copy another person's work. You will take a quiz in lab (DH 448) each week. If you've done the work and reviewed your lab manual prior to the quiz, you should have little or no problem; if you've copied another student, you will struggle. You can come to lab as many times as you need to in order to finish the lab, but you can only take the quiz once.

4. Stations should be done in sequence (i.e., Station 1, Station 2, etc.). Please do not remove materials from their original stations at the bench.

5. Clean up your lab station before moving on to the next. There are three different types of waste receptacles: 1) biohazard container; 2) glass container; 3) regular trash can. **It is important to know what waste goes in which container.** When you misplace trash, you compromise the safety of other students, our lab technicians, and the custodial staff.

6. To get credit for lab each week, you must do <u>ALL</u> of the following: 1) show your completed lab manual to the lab instructor; he or she may have you re-do sections if the answers are incorrect or incomplete; 2) have lab instructor initial your lab (gray box on first page of each lab) and your name on the master roster; and 3) take a quiz at one of the computer quiz stations. If all stations are busy, put your name on a waiting list. Stay at your computer in DH 448 or in the hallway until a quiz station becomes available; do not stand behind students at the quiz station. <u>The quiz waiting list closes 30 minutes before the end of lab each day</u>. You must take each week's quiz before the close of lab that week – quizzes cannot be made up.

7. Labs cannot be made up, since our tech sets them up and tears them down each week.

8. Be sure to return the lab card to the lab instructor to get your ID back! If you keep the lab card, you will receive a zero for the lab until the card is received. Cards must be returned within a week.

9. You are welcome to borrow a textbook with a second form of ID while you're in the lab.

10. Your exams will be available for review in the Bio 21 lab. You are encouraged to look at your exam and ask the lab instructor questions. You may not keep your exam answer sheet or the exam answer key – all exams must stay in DH 448 at the lab instructor station.

PART 2: ETIQUETTE AND RULES

1. Lab is for students enrolled in Bio 21 only. Please do not bring family and/or friends with you. <u>If you do, your lab privileges may be suspended for the week or the entire semester</u>.

2. You may use the computers for Bio 21 work only – please, no web surfing, email, or using the computers for non-Bio 21 coursework.

3. We encourage you to be social learners here. However, please be considerate of your classmates – if you're a loud talker, bring it down a notch, etc. Look for more intelligent and creative outlets to profanity – yes, some people are offended.

4. <u>Turn your cell phones off or to vibrate</u>. If you are expecting a call, take the call out in the hallway. Do not text message while in the lab – others could be waiting for a seat in lab, so be considerate and work productively.

5. Abuse the equipment and you will be told to leave.

6. Absolutely no food or drink at stations or computers! Leave food/drink on the gray cabinet by the classroom phone if you bring them to class.

7. The Department of Biological Sciences reserves the right to revoke lab privileges at any time during the semester for inappropriate or offensive behavior. If a student loses his or her privileges, he or she will not receive credit for lab(s) missed.

We greatly appreciate your cooperation in making the Bio 21 lab a unique and productive place to learn. Please don't hesitate to contact the Bio 21 faculty if you have any questions or concerns. Thank you!

PART 3: INFORMATION ABOUT THE LAB MANUAL

Your lab manual contains a number of different exercises for each week. These include:

> ACTIVITIES: These are denoted by a box outline, such as what you see here. Activities may consist of internet exercises, workbook style exercises, reading exercises and/or multimedia exercises. Most of the time, you can do the activities at home or outside of lab. The exception to this is activities requiring multimedia software and videos that are available only in lab.

Time Management Tip: Review your lab manual before coming to lab! Determine what you can or want to do outside of lab. The busiest time in lab is immediately following lecture and on the last day lab is open each week. If you want to avoid standing in line at the stations or waiting for a computer, pick another time.

STATIONS: These are exercises that are done in lab, usually on the "bench" (that is, the countertops along the perimeter of the lab). Stations are usually (but not always!) set up clockwise around the room. Please note that some stations require a partner—make friends early and often in Bio 21 so you have someone to work with.

PART 4: GET STARTED!

You will now complete your first lab for the class. There are five stations – be sure to do them in order (Station 4 is an oral quiz with the lab instructor; you must do this before going on to Station 5). When you have finished all the stations, be sure to show your lab manual to the lab instructor to have him or her initial the lab and the master roster. Review the lab and take the on-line quiz at the quiz station by the door. Good luck!

STATION 1: A MICROSCOPE PRIMER

Many weeks we will be showing you different microscopic views of the human body – from bone cells to the organisms that cause syphilis. Spend a few minutes learning how to use the microscope so you can fully appreciate this detail.

Using the diagrams and books at this station, identify the parts of the compound light microscope shown on the following page.

- Eyepiece or ocular: this is what you look through.

- Coarse adjustment knob. Used to *initially* focus on an object viewed under low power.

- Fine adjustment knob. Used to focus on an object viewed under high power.

- Low power objectives (lenses): magnifies an image at low resolution; usually the shorter lenses. Always look at the side of the objective to see what the power (i.e. magnification) is. On the Leica CME, there are two low power objectives (4x and 10x).

- Medium and high power (dry) objectives (lenses): magnifies an image at medium and high resolution. On the Leica CME, "medium" power is 40x and high power is 100x. You don't need to memorize this since you can always look on the side of the objective.

- The stage control knob moves the slide.

- The diaphragm controls the amount of light reaching the slide and specimen.

- The light switch brings power to the world of cells...

Human Biology: Lab 0

LEICA CME

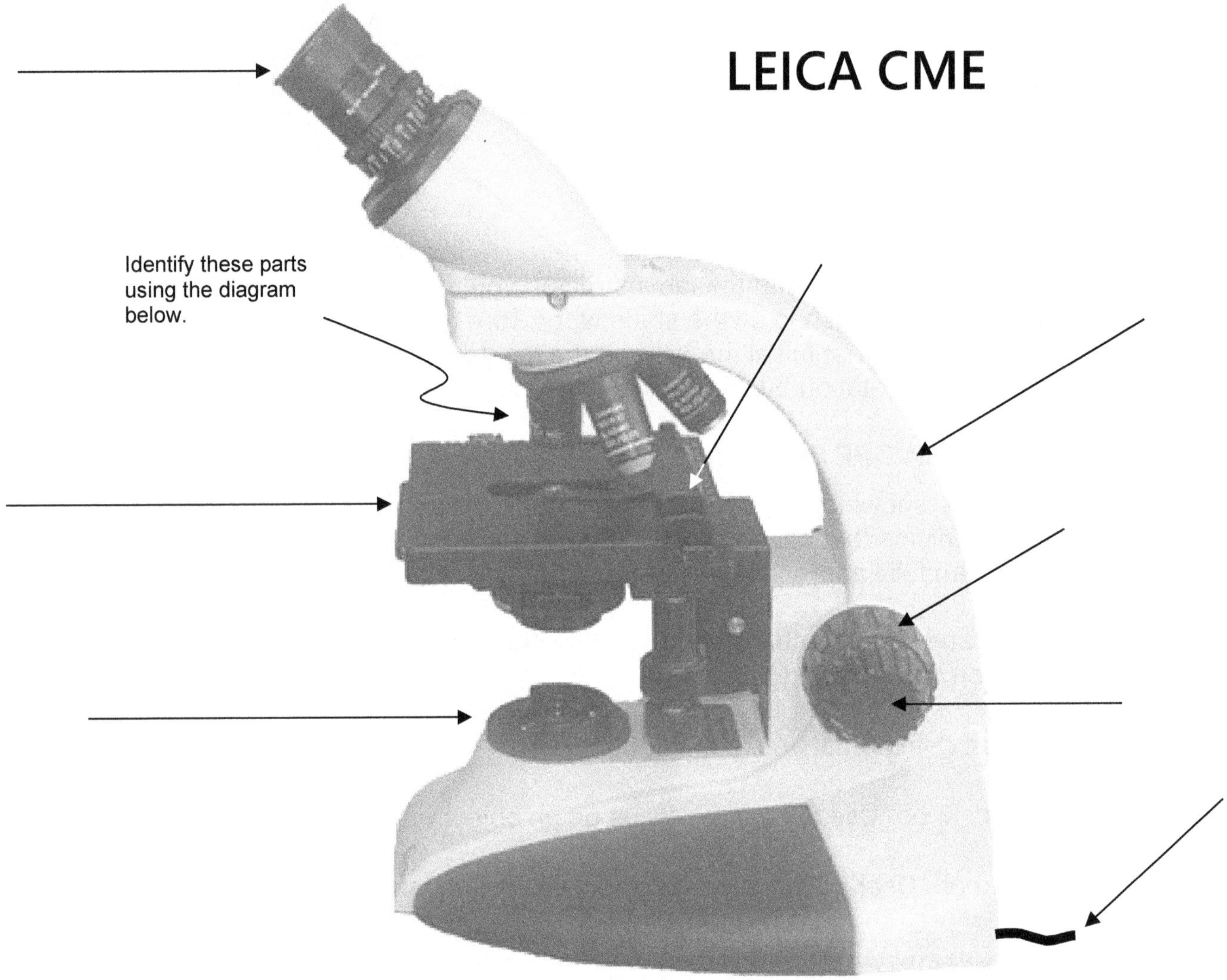
Identify these parts using the diagram below.

LEICA
4x/0.1

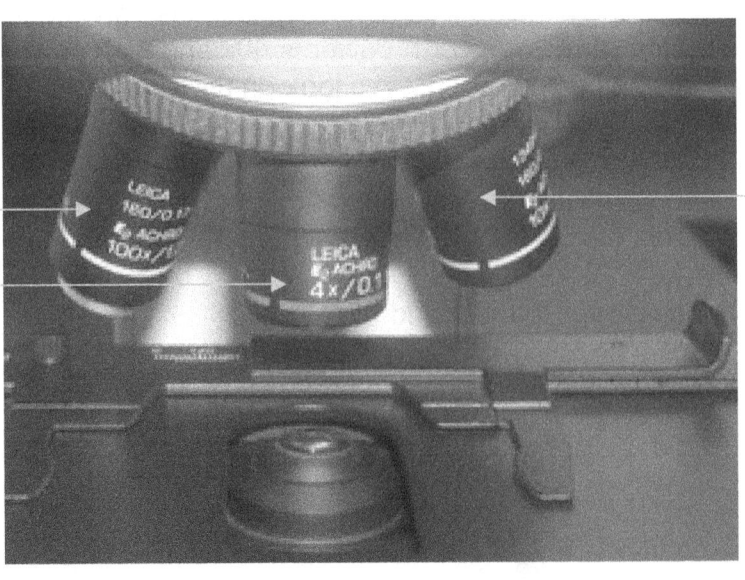

STATION 2: YOU'VE GOT THE POWER!

To get total magnification of a particular view, multiply the ocular lens (the eyepiece, usually 10x) by the objective lens (e.g., 43). So if the ocular is 10x and the objective lens is 43x, the total magnification is 430x (or 430 times the actual size of the specimen on the slide.)

On the Leica CME microscope:

1. What is the total magnification of something viewed under the lowest power at Station 2?

_____

2. What is the total magnification of something viewed under high power at Station 2?

_____

STATION 3: TAPPING INTO A WHOLE NEW WORLD

Here's the secret to viewing slides under the microscopes in this class:

ALWAYS START ON THE LOWEST MAGNIFICATION POWER

If for no other reason, this will help you get the slide's contents under the lens so that you can bring the slide into focus. Use the stage controls to move the slide up/down/left/right to bring the slide contents under the lens. Sometimes it's helpful to first do a visual with the "naked eye" (meaning, do not look down the microscope through the ocular lens – just look at the exterior of the microscope to get an idea of the position of the slide relative to the lens).

Once the slide is positioned under the lens, bring the specimen(s) into focus using the **COARSE ADJUSTMENT KNOB**. Once you have something in view, then, and only then, can you switch to **HIGHER MAGNIFICATION** and use the **FINE FOCUS ADJUSTMENT KNOB**. NEVER USE COARSE ADJUSTMENT KNOB WITH THE HIGH POWER OBJECTIVE. **You can break the slide or the lens if you do this.**

Spend a few minutes looking at a slide of human skin under the microscope. Work first with the coarse adjustment knob on low power, and then the fine adjustment knob on high power.

There are no questions to answer at this station, but you should challenge yourself by taking something completely out of focus (e.g., take the slide off the stage, move the stage, crank the coarse adjustment up and away from the stage) and then bring it back into focus. You will need to do this for your quiz at the next station!

STATION 4: MICROSCOPE QUIZ!

When you're satisfied that you know the parts of the microscope (and their functions) and how to bring a slide into focus, the lab assistant will quiz you at this station.

You must take and pass the quiz before proceeding with the rest of the lab.
You must have the lab instructor initial this box after you do this.

Human Biology: Lab 0

STATION 5: MORE ABOUT THE MICROSCOPE

At the first microscope at this station, you will find the letter "**e**" mounted on a slide.

3. Look through the eyepiece and turn the stage control knob counter-clockwise to move the slide towards the left side of the stage. What direction does the letter "**e**" move? _____

4. Look through the eyepiece and turn the other stage control knob to move the slide away from you (towards the wall). What direction does the letter "**e**" move? _____

Hint: if you are having trouble with this, look back at the first paragraph under information for Station 3.

At the next microscope at this station, you will find three colored threads mounted one on top of the other. Things mounted on slides are not of equal thickness in all places. The depth to which your microscope can focus (i.e., the depth of field) is very limited. When you use the high power objective, your depth of field is very shallow. Get around this by using the fine adjustment knob to progressively bring other parts of the slide into focus.

Note: A common mistake of people just learning the microscope is to focus on dust and debris that covers the cover slip of a slide; if you see this, keep focusing with the fine adjustment knob.

Use the microscope to determine which color thread is on top, and which is on the bottom. You may have to adjust the light using the diaphragm.

5a. Which thread is on top? _____   5b. Are you at microscope A or B? _____

6. Which thread is on the bottom? _____   Hint: in order to discern which thread is on top, first open the condenser aperture diaphragm fully (that is, with the light as bright as possible) then, using the fine focus, move the plane of focus up and down so that the colored threads come into focus one at a time.

Final Steps to Finish the Lab:

- Show your lab manual to the Lab Instructor and have him or her initial it, as well as initial the master roster.

- Review the information in this lab and take the on-line quiz.

Note: Because the Bio 21 lab is not available on a drop-in basis during the first two weeks of the semester (when we are doing orientation), you will need to take the quiz before you leave lab today. Ordinarily, however, you can take the quiz any time during the week. Just remember: the quiz for each lab must be taken the week that particular lab is assigned. Do not put off taking the quiz until the last day of the week, as you may encounter a significant waiting list for the quiz computers and the lab will close promptly at the posted time, whether or not everyone on the waiting list has taken the quiz. Planning ahead is key to your success in Bio 21.

BIOLOGY 21
LAB 1: CELLS & TISSUES

This week we take on the world of cells. The cell is the most fundamental unit of life. Your eyes, your skin, your blood, your brain, your heart – they are all made up of cells. Obviously these parts of your body are different – they look different and have different roles. We see this at the level of the cell: brain cells look different from blood cells, for example. When first learning about cells, we start with a "generalized" cell.

ACTIVITY 1: IDENTIFYING PARTS AND FUNCTIONS OF THE CELL

Each cell is, to a certain degree, a little microcosm enclosed inside a cell membrane. See Figure 4.1 in the textbook. This is a diagram of a "general" cell that, for teaching purposes, shows all the structures that appear in different cells of our body. In reality, the structures found in specific cell types (such as cells that line our digestive tracts) will have some but not all of the structures found in this diagram. Answer the following questions using information presented in Chapter 4 (see also `askabiologist.asu.edu/content/cell-parts`).

1. Name four organelles and their functions:

   a. _____

   b. _____

   c. _____

   d. _____

2. Name the three main components of the plasma membrane:

   a. _____ b. _____ c. _____

3. Describe two major ways molecules and ions cross the plasma membrane. _____
   _____
   _____

4. Briefly describe the four steps in the production of ATP from glucose, including where each step occurs. _____
   _____
   _____
   _____

Human Biology: Lab 1

ACTIVITY 2: ISSUES WITH TISSUES...

If we were amoebas, our study of our inner workings would now be concluded. However, we are complex multicellular organisms and so our studies continue...While being a multicellular organism is much more complicated than being a unicellular organism, multicellularity has its advantages. We can grow bigger (and are therefore food for fewer other organisms), and we can organize our cells to be able to better sense and react to our environment. Our cells have banded together into specialized groups of cells carrying out specific functions. These different groups of cells are tissues. See Chapter 5 in your textbook to answer the following.

5. What are tissues? _____

   _____

6. List the four types of tissues and give an example of each.

   a. _____

   b. _____

   c. _____

   d. _____

7. Where in the body do you find the following types of epithelium?

   a. simple squamous epithelium: _____

   b. simple columnar epithelium: _____

   c. stratified squamous epithelium: _____

STATION 1: LOOKING AT A FEW CELLS OF YOUR OWN...

Here you will take a sample of stratified squamous epithelium from inside your mouth and look at it under the microscope. We will use a blue stain which will make the cheek cells stand out better under the microscope. THE KEY TO THIS ACTIVITY IS TO USE A VERY SMALL AMOUNT OF DYE!

a. Put the following materials on a KimWipe™ tissue: 1) a clean toothpick; 2) a clean microscope slide and 3) a clean cover slip (the little squares; be careful – these are glass)

MORE▶

b. Using the clean toothpick, gently scrape the cheek surface on the inside of your mouth.

c. Spread your saliva (and the cheek cells in it) from your toothpick onto the center of a microscope slide. Let the saliva dry on the slide (a minute or so).

d. Put the slide on a KimWipe™ on the counter. Place a SINGLE, SMALL drop of dye on the area of the slide with your saliva.

e. Gently place the cover slip on the slide.

f. Touch a KimWipe to the edge of the cover slip to soak up excess dye, and then place the slide on the stage of the microscope.

g. Starting on the LOWEST POWER, use the coarse adjustment knob to bring the cells into view (they will appear as little blue blobs). Then use the fine adjustment knob to focus. Once you see cells, increase the magnification to MEDIUM POWER. On the right, draw what you see. (Hint: to tell if you have a cell or not, look for the nucleus – it should stain dark blue. A common error is to mistake air bubbles for cells!)

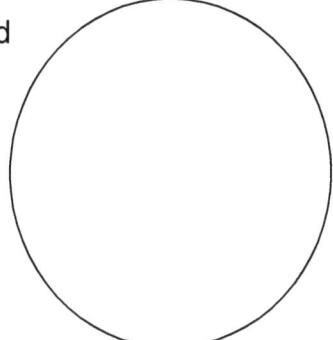

h. When you are done admiring your cells, put the slide and cover slip into the <u>glass disposal</u> box. Place your toothpick in the <u>biohazard</u> bin. <u>Wash your hands to get rid of any blue dye</u>.

8. What are two distinguishing features of the cells you saw under the microscope?

a. _____ b. _____

ACTIVITY 3: ISSUES WITH TISSUES CONTINUED: CONNECTIVE TISSUES

Refer to page 78 in your textbook to answer the following questions.

9. Name four major types of <u>specialized connective tissue</u> found in the body.

a. _____ b. _____

c. _____ d. _____

10. Name 2 cells found in specialized connective tissue (see Figures 5.3–5.6 in textbook)

a. Type of connective tissue: _____ Cell _____

b. Type of connective tissue: _____ Cell _____

11. Which of the cells in Figure 5.5 in the textbook are involved in protecting the body in some way (e.g., response to trauma or infection)? (Several are shown – name just two.)

a. _____ b. _____

ACTIVITY 4: Skin Cancer Awareness

This week we introduce you to webmd.com, a web-based personal health information resource. Go to `www.webmd.com`, type in **melanoma skin cancer** in the search window, and click **Search**. Scroll down past the ads.

This brings you to "Melanoma/Skin Cancer Health Center." Scroll down to "Overview" and click on "Skin Cancer Basics."

12. What are the three types of skin cancer? _____

_____

13. Of these, which two are most common? _____

   Which form is the most difficult to treat? _____

Scroll back up and click on Videos (on the left). Watch the video "Is that mole skin cancer?" (This link is also available through Bio 21's Canvas site under Lab 1.)

14. What is the "ABCDE" method of skin cancer detection?

15. Next, go to the American Cancer Society's "Be Safe in the Sun" web page through Bio 21 on Canvas or directly at `www.cancer.org/Healthy/BeSafeintheSun/index` Go to "Take Steps to Protect Yourself" and then to "How Do I Protect Myself from UV Rays?". (This link is also on Bio 21's Canvas site under Lab 1.) What five things can you do to protect yourself in the sun?

16. Finally, take the American Cancer Society's Sun-Safety Quiz at `www.cancer.org/healthy/be-safe-in-sun/sun-safety.html` (This link is also on the Bio 21 Canvas site under Lab 1.)

Remember to take the quiz for this week's lab exercise in the Bio 21 lab!!!

BIOLOGY 21
LAB 2: ENZYMES & DIGESTION

NOTE: THIS LAB _REQUIRES_ A PARTNER FOR STATIONS 1 & 2!

Food is vital for life. As you read this, your brain cells are consuming glucose – a very simple sugar that is the sole source of energy for your brain! Glucose and other small molecules also fuel cells over the rest of your body. Over the course of millions of years, a very efficient digestive system has evolved for breaking down the carbohydrates, proteins, and fats in your favorite Noah's Bagel (or whatever our ancestors were eating 4 – 5 million years ago on the African savannah) into molecules such as glucose.

It's no coincidence that we love to eat. Sensory receptors for taste and smell have co-evolved with our digestive system. Because we find food pleasurable, we are compelled to eat – nature's clever solution for ensuring that we seek out and obtain energy. Think about it: do you "feel" like eating when you have a cold? Probably not. The reason is not that your body doesn't need food – it probably does to launch a counterattack against your illness – but rather that you are unable to taste and/or smell food.

ACTIVITY 1: YOUR GUTS

Refer to Chapter 16 in your textbook to answer the following questions.

1. Diagram the pathway of food through the different organs of the digestive tract, starting with the mouth and ending the anus (do not include accessory organs – we'll come to those in a bit). Use arrows to show the path, i.e., Structure A ➔ Structure B, etc.

   _____

   _____

2. In which organ along the digestive tract are most nutrients absorbed? What specializations does this organ have for absorption? _____

   _____

3. What is the function of the following organs?

a. salivary glands _____

b. liver_____

   _____

c. gallbladder _____

d. pancreas_____

Human Biology: Lab 2   2.1

STATION 1: CARB DIGESTION (Adapted from *Laboratory Manual Human Biology* by Sylvia Mader)

The macromolecules in the food we eat (carbohydrates, fats, proteins) are typically too large and complex for our bodies to use directly; they must be broken down into smaller molecules in order to be taken up and utilized. This process of breakdown can in many cases be thought of essentially as a reaction between the macromolecule and the water surrounding it in which water splits the big macromolecule into parts (smaller molecules). This reaction is called hydrolysis ("hydro" referring to water, "lysis" meaning "to split"). While this reaction *can* occur spontaneously, it is very slow. It might take weeks or months, for example, for a large starch molecule (a complex carbohydrate) to be *spontaneously* broken down into its simpler component parts (simple sugars).

In order to accelerate reactions, the cells in our bodies synthesize different kinds of protein molecules called *enzymes*. Enzymes increase the rate of reaction by binding a specific substrate (a carbohydrate molecule, for example) in such a way that it is easy for water (or some other substrate) to react with it. In the presence of the enzyme, water reacts with the carbohydrate very quickly, rapidly breaking it down (hydrolyzing it) into simple sugar molecules that the body can use. One way to diagram this reaction is presented below:

Carbohydrate + water + enzyme → simple sugars + enzyme
(substrate) + *(substrate)* + *enzyme* → *(products)* + *enzyme*

Note that the enzyme is not destroyed in this reaction.

Since enzymes are very specific for the substrates they bind to and the reactions they accelerate, our cells must synthesize many different types of enzymes. To facilitate the breakdown of complex carbohydrates, fats and proteins in our food, for example, a variety of enzymes must be used.

Think the Problem Through

Salivary amylase, which is present in your saliva, initiates the breakdown (i.e., digestion) of starches found in carbohydrates (starches) such as bread, potatoes, etc. The process can be shown as a simple equation:

starch + water $\xrightarrow{\text{amylase (enzyme)}}$ maltose (a sugar)

Put another way, starch and water, in the presence of amylase, will break down into the simpler sugar maltose. Put even more simply, if you chew a saltine cracker and mix it with your saliva (which contains water and amylase), you will end up with a wad of maltose in your mouth. Additionally, the amylase is not used up or changed during this process.

4. If digestion *does not* take place, will you have starch or maltose? _____

5. If digestion *does* take place, will you have starch or maltose? _____

In this activity, you will use a test for starch digestion. Here is a key piece of information that will help you with understanding what's going on in this activity:

Iodine is frequently used to test for the presence of starch. In this test, you add a drop of iodine to the starch mixture. <u>If the mixture turns blue-black, starch is present</u>. In other words, digestion has not taken place.

a. Start with two small beakers at Station 1. Set up the materials below on the template available at this station.

b. Pour 10 ml of starch solution (it is 1 percent starch by weight) into each beaker.

You will now add a solution containing an enzyme (bacterial amylase in this case) into the beaker labeled "Starch + Enzyme." Before you do this, predict what will happen to the starch when you add the enzyme (it might help to diagram the reaction).

6. _____

_____

_____

c. Use the medicine dropper to add about 3 drops of the enzyme solution to the beaker labeled "Starch + Enzyme". Using a clean toothpick, mix up the solution.

d. Test for the presence of starch in the beakers by doing the following: pour a few drops of the solution from the first beaker ("Starch only") <u>into Well #1 of the spotting plate</u>, add one drop of iodine and stir with a toothpick. Then pour a few drops from the second beaker ("Starch + Enzyme") into <u>Well #2</u> of the spotting plate, add one drop of iodine and stir with a toothpick. Iodine and starch will begin to spontaneously react to form a colored complex. Observe the color of each mixture and record your results in the table below in columns one and two.

Color of starch solution AFTER adding iodine

at beginning of experiment		at 5 minutes	
Well #1 (starch only)	Well #2 (starch + enzyme)	Well #3 (starch only)	Well #4 (starch + enzyme)

e. Kick back for 5 minutes. Engage your lab partner in sparkling conversation. At the end of 5 minutes, test for the presence of starch in the beakers again. Pour a few drops of the solution from the first beaker ("Starch only") into <u>Well #3</u> of the spotting plate, add one drop of iodine and stir with a toothpick. Then pour a few drops from the second beaker ("Starch + Enzyme") into <u>Well #4</u> of the spotting plate, add one drop of iodine and stir with a toothpick. Iodine and starch will begin to spontaneously react to form a colored complex. Observe the color of each mixture and record your results in the table above in columns three and four.

7a. What happens to the "starch only" solutions when you add iodine to them? _____

_____

7b. Does this change over time? _____

8a. What happens to the "Starch + Enzyme" solution when you add iodine to it? _____

_____

_____

8b. Does this change over time, and if so, what do you think is causing the change? _____

_____

_____

Please rinse and dry your beakers and spotting plates, and return them to the station. Clean up the station for the next students. Thanks!

STATION 2: ENZYMES IN FRESH VS. PROCESSED FOODS

While in nature enzymes are neither used up nor altered in a reaction, commercial food processing techniques can destroy enzymes in certain types of food. In this activity, we'll compare enzymes in fresh pineapple with canned pineapple. Fresh pineapple contains a protein digesting enzyme called bromelain. Fresh pineapple is often eaten at the end of a meal because it contains this digestive enzyme. But when pineapple is heated during the canning process, the protein structure of the bromelain is broken down (or denatured). So while the canning process is great for killing bacteria, it inactivates the enzymes present which happen to be beneficial to us. Incidentally, some good background on Jell-O™ that may help you with questions at the end of this activity can be found at
`www.howstuffworks.com/question557.htm`

What to Do
Follow directions at Station 2 for setting up your Jell-O™.

Overview: You will take a Petri dish containing a Jell-O™ sample and place it on a template with three areas demarcated (water, fresh pineapple, and canned pineapple). You will then add water, fresh pineapple or canned pineapple to these areas. After 45 minutes, you will determine whether "digestion" has occurred in each of these three areas. After you have set up your Jell-O™, move on to Question 9.

9. What is the purpose of having a section of the plate with just Jell-O™ and water? _____

_____

While your experiment is running, you can do other activities in this lab (we recommend doing Activity 2 and/or Activity 3).

At the end of 45 minutes, check the contents of the three areas in your container for enzyme activity. The Jell-O™ will appear to melt somewhat in the presence of enzymes. Record the observations in the table below. Place your Jell-O™ container in the appropriate bin as indicated in the directions on the countertop. Return the white try to the bin at the beginning of the station.

Analyzing Your Data

Solution	Observation (no melting, some melting, a lot of melting)	Enzyme activity? (y or n)
1. Water		
2. Fresh Pineapple		
3. Canned Pineapple		

10. What do you conclude about enzyme activity in fresh vs. canned pineapple? _____

_____

_____

Refer to the box of Safeway Lemon Gelatin Dessert on the counter. On the back of the box, the instructions indicate that fresh or frozen pineapple, kiwi, ginger root, papaya, figs or guava should not be mixed in with the gelatin because the gelatin will not set. Why won't the gelatin set? (You will not be able to figure this out unless you read the information on the previous page under "Station 2"!)

11. _____

_____

12. Next, on the box is a recipe for Lemon Pineapple Sparkle. Why could one be reasonably

sure this recipe will work?_____

_____

_____

Human Biology: Lab 2 2.5

ACTIVITY 2: CELLULAR RESPIRATION

Aerobic respiration:
Go to "What is Aerobic Respiration?" at www.youtube.com/watch?v=ZkqEno1r2jk (this is also linked on the Bio 21 Canvas site under Lab 2).

13. Aerobic respiration means with air and so needs _____.

14. Aerobic respiration releases energy in cells by _____
In the presence of oxygen.

15. What is the simplified equation for this?

16. Now state this equation in a full sentence.

17. What is the energy used for?

18. What is ATP essentially? (Note: we are not asking what ATP stands for, but what it is!)

19. Where does aerobic respiration typically occur?

Anaerobic respiration:
Go to "What is Anaerobic Respiration?" at www.youtube.com/watch?v=HZtXLhm7ISA (this is also linked on the Bio 21 Canvas site under Lab 2).

20. What is the general equation for anaerobic respiration?

21. Why is less energy produced in anaerobic respiration?

22. What is the problem with lactic acid as a byproduct?

ACTIVITY 3: The Virtual Body: Homeostasis

Homeostasis refers to the body's ability to maintain a constant environment, including temperature, water, blood sugar, etc. Many chemical reactions that are vital to life are possible only within certain temperature ranges. If you get too cold or too hot, for example, your body will respond by shivering to warm you up or sweating to cool you down.

For this activity, you will view the "The Virtual Body: Homeostasis" at www.youtube.com/watch?v=QKT47A-LBj4 (Also on Bio 21 Canvas site under Lab 2.) Answer the questions below. For this week's lab, you will only view the video up to about 13 minutes (we'll revisit homeostasis again in Lab 7).

23. What is "negative feedback"?

24. *Temperature.* At the cellular level, heat affects _____.

25. The hypothalamus is the body's thermostat. What exactly does it do to regulate body temperature?

26. *Blood Glucose.* Blood glucose is required for _____.

27. The pancreas has receptors that detect a rise in blood glucose. What hormone is released when blood glucose rises? _____. This hormone removes sugar from the blood in two ways: 1)_____

and 2) _____.

28. What happens if glucose levels fall too far?

We'll return to this video in Lab 7 to learn about water balance and the kidneys.

BIOLOGY 21
LAB 3: SKELETOMUSCULAR SYSTEM

NOTE: THIS LAB *REQUIRES* A PARTNER FOR STATION 5!

For animals, one of the most basic life requirements is the ability to move around. We engage our bodies in hundreds of motions each day, even as we sleep. In fact, we find it is uncomfortable to stay "still" for very long – our necks start to hurt, our feet "fall asleep", etc. Motion is the rule rather than the exception. This week we'll learn more about the body's skeletal and muscular systems.

ACTIVITY 1: OVERVIEW OF BONES

1. Bones are the main organs of the skeletal system. Refer to Chapter 6 in your textbook. What are the functions of bone and the skeletal system?

2. Places in the body where two or more bones connect are called joints. Although you are probably most familiar with joints at the shoulder and elbow, there are many others such as the wrist, hip, knee and ankle. Refer to Chapter 6 in your textbook. What are the 3 different functional types of joints?

3. The synovial joint is the most common type. Refer to pages 111-113 in your textbook. What kinds of movement are possible with a synovial joint? Name two synovial joints in the human body.

STATION 1: SPENDING TIME WITH ELVIS

A. Bone Identification. Pull up a chair and spend a few minutes with Elvis, the resident human skeleton in the Bio 21 lab. Using your textbook (Chapter 6) and/or charts at this station, compare the bones in your diagram with Elvis' bones. Identify the different bones of the skeleton (on your lab worksheet, not on Elvis). See next page for worksheet.

Human Biology: Lab 3   3.1

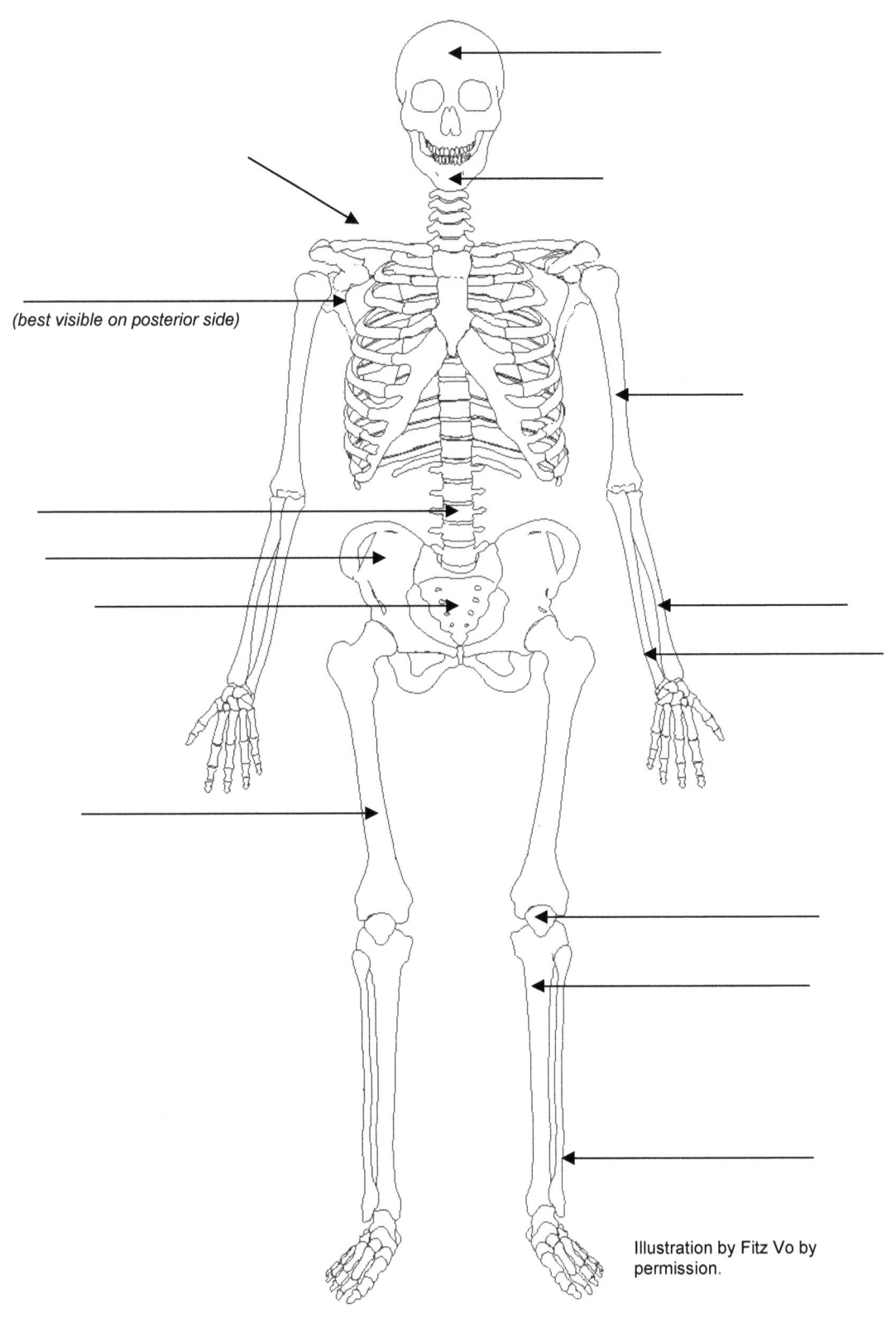

(best visible on posterior side)

Illustration by Fitz Vo by permission.

B. Surface Features. Next, we'll consider some features of bones you can feel on your skin. As you try to identify some of these bones, use Elvis for comparison.

> Terminology Hint: a "process" is the scientific name for a projection or bump on a bone.

4a. Name the process (i.e., a projection on a bone) that is found on the elbow joint _____.

4b. Touch this process and alternately extend and flex your forearm. Which part of the arm does this process move with: the forearm or the upper arm? _____.

4c. While still touching this process, alternately turn the palm of your hand over and back (*gently* try this with Elvis). Does this process move a lot? _____ What bone is this process on? _____.

Identify the bones to which the surface features listed in the following table belong. Refer to the diagrams and charts at Station 1.

Surface Feature	Bone (hint: think about long bones!)
Knuckles	
Bone with bump next to the wrist and on the same side your little finger (pinky)	
Bone with smaller bump next to the wrist and on the same side as your thumb	
Bone with bump next to and outside the ankle	
Bone with bump next to and inside the ankle.	

STATION 2: EXTERNAL ANATOMY OF BONE

Note: this material is fragile! Please handle it very carefully!

Look at the femur (i.e., thigh bone), the longest bone in the human skeleton. It consists of a shaft, or diaphysis, with two knobby ends or epiphyses (singular: epiphysis).

5a. On one end, this bone has a narrow neck and a round head. What bone does this head articulate (or join) with? _____

5b. Compare this joint with the one formed between the humerus and ulna. How are they different? (Hint: think about differences in range of motion...) _____

5c. What feature of the head of the femur allows its wide range of motion? _____

5d. The other end of the femur is broad with two flat surfaces that articulate (or join) with another bone called the _____

5e. What is the range of motion of the joint formed between these two bones? _____
_____

5f. What makes this range of motion possible? _____

5g. Why are knees so vulnerable to injury (usually in sports)? (Hint: think about how this joint differs from the hip and elbow joints) _____

Finally, note some of the features on the femur, such as lines and projections of various sizes. These surface features are attachment sites for tendons and ligaments.

STATION 3: INTERNAL ANATOMY OF BONE

Look at the inside of the femur at this station. There are two kinds of adult bone tissue: compact bone and spongy bone. Compact bone is solid and dense, and is found on the surface of the femur. Spongy bone is lattice-like and is found on the inside of the femur, primarily in the epiphyses and around the marrow cavity.

Refer to the bone on display at this station and the illustration on the right to help you answer the following:

6a. What type of bone is found in the shaft of the long bone (i.e., the diaphysis)?

6b. What are some of the properties of this type of bone?

7a. What type of bone is found inside the ends of the long bone (i.e., the epiphyses)?

7b. What are some of the properties of this type of bone?

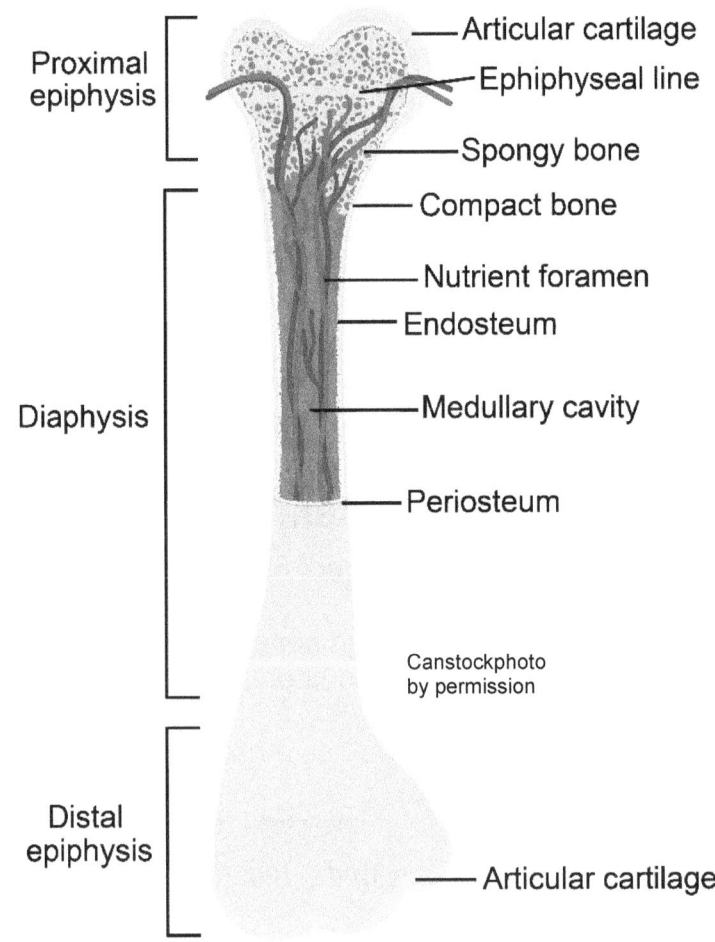

Human Biology: Lab 3

STATION 4: BONE UNDER THE MICROSCOPE
Using only the fine adjustment knob on the microscope, look at the slide of compact bone under the microscope. This slide has been produced by grinding a piece of the shaft of a long bone with coarse and then finer stones until a thin wafer remains. Although only the mineral part of the matrix is present, the basic architecture is preserved. In living bone, blood vessels and nerves are present in the large central canals. These are surrounded by concentric layers of matrix called lamellae (singular: lamella). There are intervening rings of lacunae (smaller holes) in which living bone contains cells called osteocytes. Find a central canal and identify lamellae and lacunae. Compare what you see with the model next to the microscope. Draw the different structures you see under the microscope and label the following parts: central canal, lacuna, and lamella.

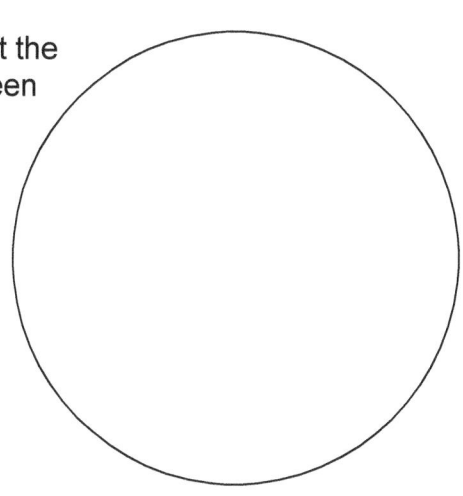

ACTIVITY 2: MUSCLES

Skeletal muscles are the main organs of the muscular system and are composed primarily of skeletal muscle tissue. Skeletal muscles are connected to bones by dense fibrous connective tissue called tendons. When a skeletal muscle contracts, movement may or may not occur. If the skeletal muscle is allowed to shorten, the bone moves as does some part of the body. On the other hand, if the skeletal muscle does not shorten, the tension in that muscle and its tendons increases. All skeletal muscles exhibit tension or muscle tone. This tension maintains the position of the body against the pull of gravity. The ability to hold the body erect is called posture. What follows next is a somewhat deeper look into how muscle actually contracts. Refer to Chapter 6 in your textbook to answer the following:

8. What are the two general purposes of muscles (hint: look at section headings in Chapter 6)

a. _____

b. _____

9. How is skeletal muscle different from smooth and cardiac muscle? _____

_____

_____

10. Many skeletal muscles occur as "antagonistic" pairs. This means that in order for movement to occur, one muscle must relax while the other contracts. See p. 115 in textbook and give an example of one such pair, describing what the motion is, what muscle(s) are contracting and what muscle is relaxing.

_____

_____

Human Biology: Lab 3

11. A single *skeletal* muscle is composed of many bundles of muscle cells.

a. Skeletal muscle consists of many _____ of muscle cells.

b. A muscle cell consists of many _____.

c. Two types of myofilaments are _____ and _____.

d. What is a sarcomere? _____

e. Draw a sarcomere below and label the different types of myofilaments:

STATION 5: MUSCLE FATIGUE (YOU WILL NEED A PARTNER FOR THIS ACTIVITY)

Muscles that are continually stimulated will eventually get tired. You can't hold a barbell over your head forever! The following experiment will demonstrate muscle fatigue.

a. Hold a rubber ball in your dominant hand (e.g., the hand you write with) and squeeze it as rapidly as possible. Your partner should record the number of contractions for every one minute interval for a total of 3 minutes. Record the data in the table below under Trial 1.

b. Rest for one minute (use a timer) and then repeat the contractions (same hand) for another 3 minutes (Trial 2).

c. Rest for one minute (use a timer), and again repeat the contractions (same hand) (Trial 3).

d. The partner who acted as the timer should now carry out steps a-c.

e. Answer the following questions based on the data in your table.

12. Did you experience fatigue in the 2nd trial? _____ 3rd trial? _____

13. How are your results different from your partner's? _____
_____

		Number of repetitions
Trial 1	First Minute	
	Second Minute	
	Third Minute	
	Rest one minute!	
Trial 2	First Minute	
	Second Minute	
	Third Minute	
	Rest one minute!	
Trial 3	First Minute	
	Second Minute	
	Third Minute	

ACTIVITY 3: OSTEOPOROSIS ON THE WEB

Go to www.WebMD.com, and type **osteoporosis** in the search window. After scrolling past the ads, click on Osteoporosis to get to the Osteoporosis Health Center. Answer the following:

14. What is osteoporosis and what are the causes of osteoporosis?

15. Who is at risk for osteoporosis? (Think about age, lifestyle, etc.).

16. How can you prevent osteoporosis from developing? What advice would you give to your family members (especially mothers, aunts, grandmothers, etc.)? You will need to surf WebMD (or another reputable site) for this information.

ACTIVITY 4: CALCIUM ON THE WEB

Good sources of calcium are typically found in dairy foods (good for your bones). However, many dairy foods are also high in saturated fat (bad for your heart). The key is to include in your diet a wide variety of foods that are calcium-rich but unsaturated in fat. Go to the USDA's Food Composition Database at `ndb.nal.usda.gov/ndb/search/list` (also on the Bio 21 Canvas site under Lab 3) to research both dairy-based and plant-based calcium-rich foods you can incorporate into your diet.

When using the USDA's website: Under **Select Source**, choose "Standard Reference", and enter your food item under **Enter one or more terms**. This way you will see both calcium and "fatty acids, total saturated" (i.e. saturated fats), *and* you can enter quantities in ounces.

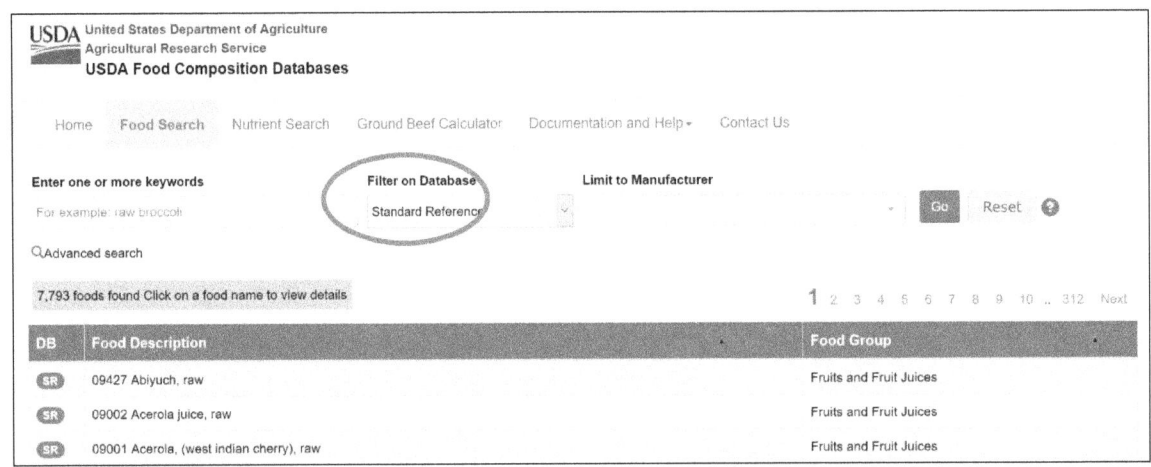

	Source (Food)	Serving Size	Milligrams of calcium per serving	Grams of saturated fat per serving
Dairy foods				
Non-dairy foods				

17. How much calcium should you be getting? Give the Recommended Daily Allowance (RDA) for your age group (if you are pregnant or nursing, be sure to give that number). My calcium requirement is _____. (Hint: use Google to find this information!)

BIOLOGY 21
LAB 4: NERVOUS SYSTEM & SENSORY RECEPTION

NOTE: THIS LAB *REQUIRES* A PARTNER FOR STATIONS 2, 4, 5 & 6!

This week, we will examine the nervous system and sensory reception. The nervous system is responsible for **emotions**, **cognition** (thinking), **movement** and **sensations** (awareness of a stimulus). To understand how the nervous system operates, you must know some basic anatomy of sensory organs and some neuroanatomy (anatomy of the brain and spinal cord).

ACTIVITY 1: ORGANIZATION OF THE NERVOUS SYSTEM

Refer to Chapter 7 in your textbook. Pay close attention to the organization of the Nervous System – the distinction between the Central Nervous System and the Peripheral Nervous System is important! (Table 7.1 will help you.) Answer the questions below.

1. What are the two divisions of the nervous system?

a. _____ b. _____

2. What are the two key anatomical structures of the **central** nervous system?

a. _____ b. _____

3. What are the two main divisions of the **peripheral** nervous system?

a. _____ b. _____

4. What are the two divisions of the **autonomic** division of the peripheral nervous system?

a. _____ b. _____

STATION 1: THE BRAIN UP CLOSE

At this station, you'll see an actual human brain (preserved), as well as models and figures illustrating the major parts of the brain. Using the models and figures, label the different parts of the brain in the diagram on the right.

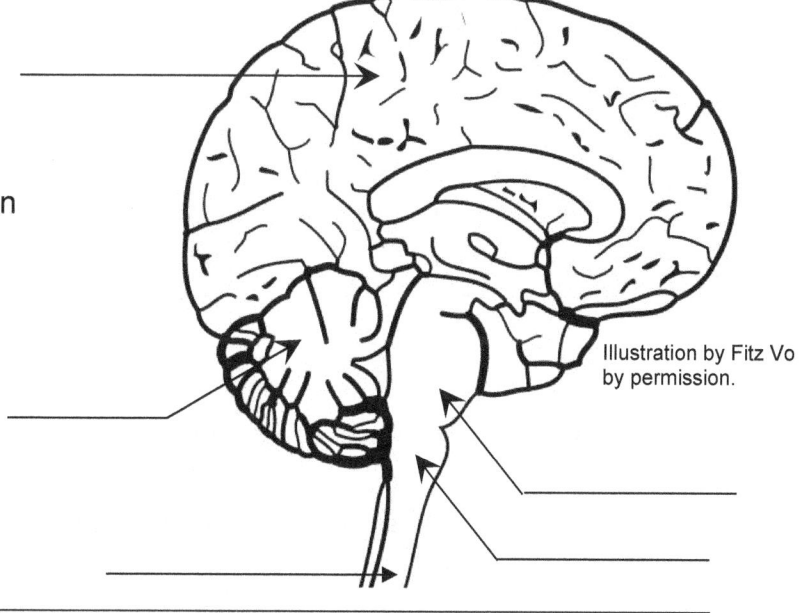

Illustration by Fitz Vo by permission.

Human Biology: Lab 4

Using information in Chapter 7 of your textbook and `healthline.com`, identify the functions of the following parts:

Part	Function
5. Cerebrum	
6. Cerebellum	
7. Corpus Callosum	
8. Hypothalamus	
9. Thalamus	

STATION 2: REFLEXES & REACTION TIME

Reflexes are unconscious responses to a stimulus. The response is stereotyped (always the same) and involuntary (without awareness). The response does not require information from higher brain centers (that is, information from the cerebrum). When a stimulus is received (for example, you accidentally cut your finger), a reflex action occurs (you pull your hand back).

Similarly, when a doctor taps your knee, your thigh muscle contacts and your foot moves forward. The tapping of the knee activates a neuron in your knee that sends information to the spinal cord. The first neuron connects to another neuron in the spinal cord, which sends information to your thigh muscle. This information tells your thigh muscle to contract. As a result, your foot moves forward. This is known as the stretch reflex. There is an additional reflex that occurs, which makes your foot return to its initial position. This is an inhibitory reflex. We will not examine the inhibitory reflex in this lab.

THE STRETCH REFLEX

a. Have your partner sit on the chair at this station, legs uncrossed and hanging loosely.

b. With the reflex hammer, gently tap the area just below the kneecap (the patellar tendon).

c. Your partner's foot should swing forward involuntarily. If it does not, you may not have tapped the patellar tendon. Try again. Ask the lab assistant for help if necessary

Answer the following questions about the stretch reflex:

10. In the stretch reflex, what is the stimulus and what is the response?

    Stimulus: _____

    Response: _____

11. Is this reflex voluntary or involuntary? _____

REACTION TIME

This fun demonstration allows you to appreciate the speed of your nervous system. During this exercise, think about the amount of time it takes your cerebrum to send information to the spinal cord and then on to your hand...*Wow that's fast!*

a. The subject rests his or her forearm on the table with the wrist at the edge. The tester will then position the ruler so that the subject's pointer (index) finger and thumb are in a "ready position" to grab the ruler. The pointer finger and thumb should be about 2 inches apart and aligned with the "*thumb-line*" of the ruler. (See picture.)

b. <u>Without any forewarning</u>, the tester will release the ruler and the subject will try to catch it.

c. Using the information on the ruler, record the reaction time below. Note the scale is in *milli*seconds (4 *milli*seconds=.004 seconds).

E. McGee by permission

Your Reaction Time (in milliseconds)	
Your Partner's Reaction Time (in milliseconds)	

d. Now reverse roles and repeat. Record your partner's data next to yours in the table above.

Answer the following questions about reaction time:

12. What is the type of nerve (i.e., motor or sensory) that carries the information to innervate

    the muscles in your hand to catch the ruler? _____

13. Is this reaction voluntary or involuntary? (Hint: think about whether catching a ball thrown

    to you without warning is voluntary or involuntary...) _____

STATION 3: EYES & EARS

At this station, you'll see models of ears and eyes. Identify the different parts of the eye and ear indicated on the diagrams, and then list their functions (refer to Chapter 8 in your textbook if you need help).

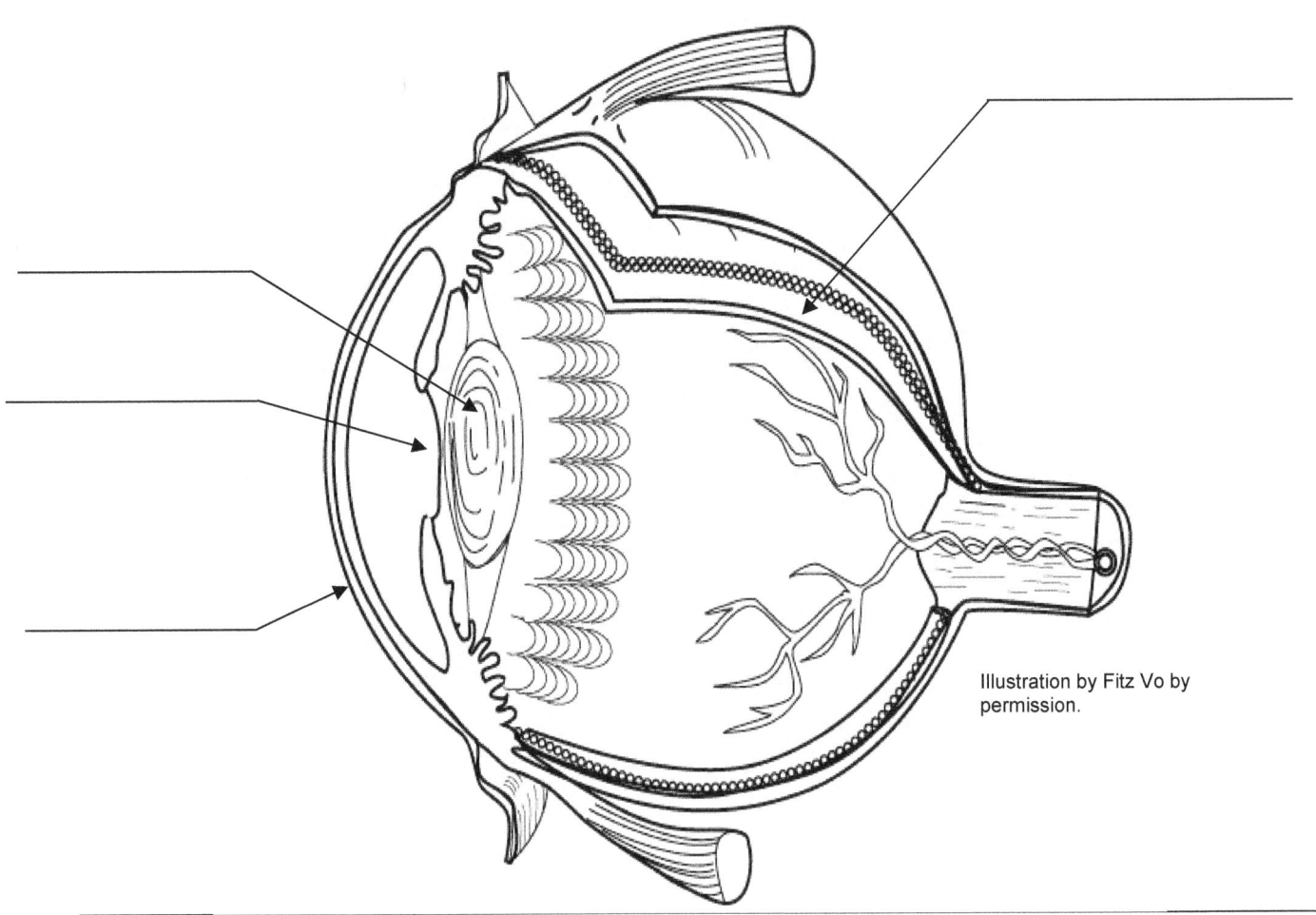

Illustration by Fitz Vo by permission.

Part of the Eye	Function
14. Cornea	
15. Retina	
16. Lens	
17. Pupil	
18. Fovea (not shown on the diagram above)	

19. What part of the eye is affected by "Lasik" surgery? (go to Google for help if necessary!)

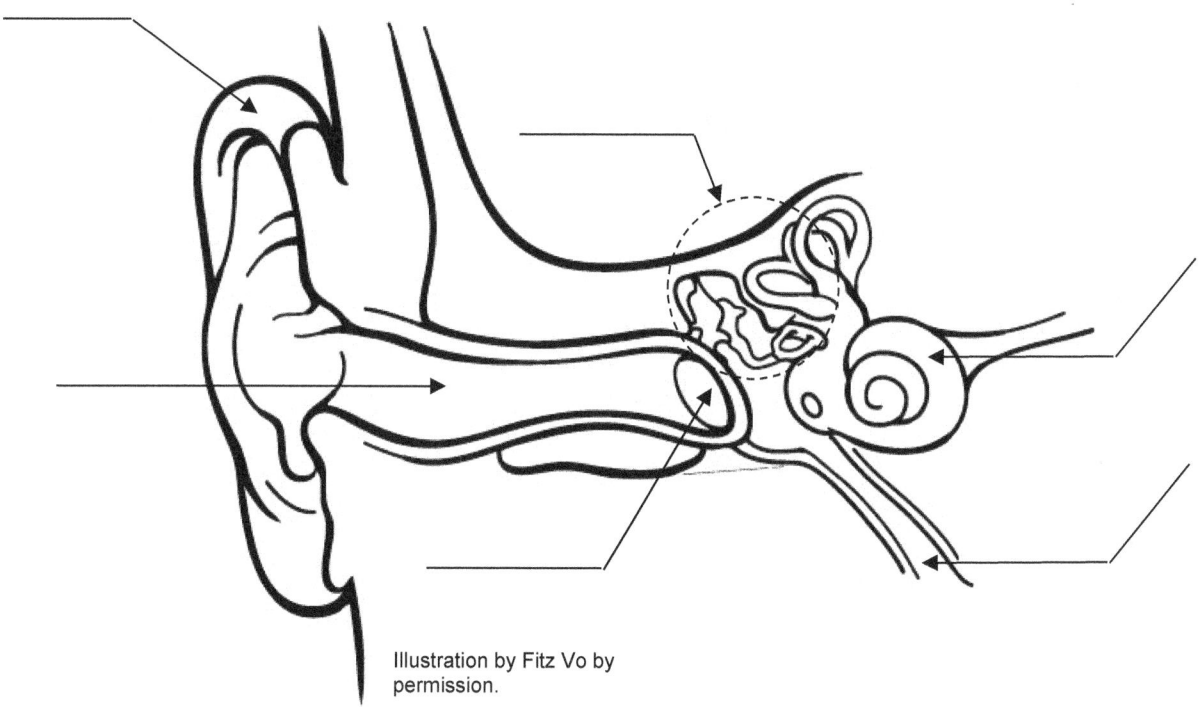

Illustration by Fitz Vo by permission.

Parts of the Ear	Function
20. Pinna	
21. Tympanic Membrane	
22. Malleus, Incus, and Stapes	
23. Cochlea	
24. Auditory Canal	
25. Eustachian Tube	

26. What is the scientific name for the ear drum? _____

STATION 4: PUPILLARY REFLEX

The <u>pupillary reflex</u> is a reflex that causes your pupils (the opening of the eye) to dilate or constrict to accommodate the amount of light that enters the eye. Again, this reflex is involuntary

a. Shine the penlight into one of your partner's eyes and observe the response.

b. Turn the penlight off and observe the response.

c. Ask your partner if he or she was aware of the pupil's response.

d. Now reverse roles and repeat.

Answer the following questions about the pupillary reflex.

27. When the light was shined in your partner's eyes, did the pupil get larger or smaller?

28. When you turned off the light (after shining it in the subject's eyes), did your partner's pupil get larger or smaller?

29. Was your partner aware of the pupil changing size?

30. Why does the pupil size change with the amount of light?

STATION 5: WHAT BIG EARS YOU HAVE!

Sound waves are conducted through the outer ear (the pinna and external auditory meatus) to the tympanic membrane (eardrum), causing it to vibrate. The movement of the tympanic membrane causes the three ossicles of the middle ear (the malleus, incus, and stapes) to vibrate one after another until the stapes pushes against a flexible membrane called the oval window. The oval window vibrates, producing compression waves in the cochlea (inner ear).

There are two general kinds of deafness (loss of hearing): 1) deafness due to damage to the middle ear (conduction or conductive deafness; usually due to infections of the middle ear or tympanic membrane, or excessive accumulation of ear wax); and 2) deafness due to damage to the inner ear (sensory or sensorneural deafness; usually the result of infections, or <u>prolonged exposure to loud sounds</u>).

At this station, you'll conduct a test to detect hearing loss.

Rinne Test (Conductive Hearing)

If you are wearing earrings, remove them before doing this activity (if you are wearing "posts", you may leave them in but take care not to put the tuning fork on any part of the earring).

a. Strike a 256 Hz tuning fork with a rubber mallet to produce vibrations. Important: do not ever strike the tuning fork against a metallic object (including jewelry)!

b. Place the handle of the vibrating tuning fork first against the mastoid process (the bony bump behind the ear), with the tuning fork pointed down and behind the ear. Then move the tuning fork (by the handle) near the external auditory meatus. Ask your partner, "Which is loudest: ONE (on the mastoid) or TWO (near the auditory meatus).

Normal hearing patients report that TWO is louder. If a person has conductive hearing loss, the opposite is true: ONE is louder.

c. Now simulate conductive deafness by putting a plug of cotton in your ear, and repeating steps 1 and 2.

31. Do your results suggest you have "normal" conductive hearing? _____

32. Did you notice a difference when you simulated conductive deafness? _____

33. How do you think hearing aids work? (Get this info from the internet). _____

_____

_____

_____

STATION 6: SPINAL NERVES, EYES AND EARS WORKING TOGETHER

Our sense of balance relies on information (cues) from the ears and the eyes. The cerebellum uses this information to help us keep our balance. If both of these cues are missing, we have difficulty keeping our balance. Try the following:

Use a timer to record accurate time.

a. Stand with both feet together.

b. Lift one foot 6 inches off the floor, bending your knee at a 45-degree angle. If you're right-handed, lift your left leg. If you're left-handed, lift your right leg. Don't move your foot.

c. Have your partner start the timer. Record in the table below how long you can maintain your balance.

d. Now reverse roles and repeat.

Now try this:

a. Stand with both feet together.

b. *Close your eyes* and lift one foot 6 inches off the floor, bend your knee at a 45-degree angle. If you're right-handed, lift your left leg. If you're left-handed, lift your right leg.

c. Have your partner start the timer. Record how long you can maintain your balance in the table below.

d. Now reverse roles and repeat.

Note: The older you are, the harder it will be for you to keep your balance because information from the neurons travels at slower rate with age.

	Balance Time
With Eyes Open	
With Eyes Closed	

STATION 7: DISTRIBUTION OF TASTE BUDS ON THE TONGUE

In order to taste food, the food's chemicals are first dissolved in saliva (spit) and then diffused into the pores of your taste buds. There are 4 types of taste cells: sour, sweet, bitter, and salty (nb: some scientists now claim a fifth taste cell, savory, exists!). Try the following activity to locate where these taste buds are on the tongue.

a. Using a **clean** applicator stick, select one of the solutions and touch various taste receptors on the tongue until you can identify the associated taste.

b. Repeat with various solutions and allow the subject to identify the taste of the solution (salty, sweet, etc.)

c. **Do not re-dip the applicator stick into the solutions—always use a clean stick. Put used applicators in the biohazard bucket.**

d. Using the figure on the right, identify where on the tongue the taste intensity for sweet, sour, bitter and salty is greatest.

e. Help yourself to a piece of candy to get rid of the taste of the test solutions!

MORE>

ACTIVITY 2: HEARING EDUCATION AND AWARENESS FOR ROCKERS (H.E.A.R.)

Recall the last time you went to a club or a concert. If you experienced ringing in your ears – either during or after – your ears sustained hearing damage. According to the H.E.A.R. website, "Damage from loud sound can occur from playing music, attending concerts, dance clubs, raves, using stereo earphones, playing amplified systems too loudly, or other noisy activities." And if you think the government is looking out for you while you have fun, you're mistaken. OSHA (Occupational Safety and Health Administration) only protects you on the job. (Although there is a notable exception on the horizon: the San Francisco Board of Supervisors recently passed the "Earplug Ordinance" for San Francisco clubs.) Promoters, club managers and owners, and musicians routinely determine the loudness of music at concerts and clubs. And if you think their choices are informed and wise, well, you'd be mistaken there, too. Keep in mind that the phenomenon of loud music isn't that old – maybe three or four decades. We're now finding all too often (and too late) that loud music causes hearing damage, loss, and deafness. This week we're pointing you to a website created by musicians for music fans, DJs, sound engineers, and other musicians. The following links, available on Bio 21's Canvas site under Lab 4, are from the parent site for H.E.A.R. at `www.hearnet.com/` Go to the links below and answer the accompanying questions.

`www.hearnet.com/at_risk/risk_tinnitus.shtml`

34. What is tinnitus?

35. What things aggravate tinnitus?

36. What should you do if you experience tinnitus?

`www.hearnet.com/at_risk/risk_aboutloss.shtml`

37. At a typical rock concert where the music is between 110 and 120 dB, how long can you listen (with unprotected ears) before sustaining hearing damage and/or loss?

`www.hearnet.com/at_risk/risk_at_risk1.shtml`

38. List three precautions you can take to protect your hearing at concerts and clubs?
    a.
    b.
    c.

39. List two precautions concert venues and clubs can take to protect patrons' hearing?
    a.
    b.

ACTIVITY 3: "FEAR NOT"

Refer to the article on Canvas under Lab 4 titled "Fear Not". Read the article and answer the following questions:

40. What two non-surgical approaches are being used to treat fear disorders? What is the combined effect of these two approaches?

    (Note: this is a holistic question – meaning you'll have to read through the entire article and then formulate a response to the question.)

41. What part of the brain contains fear memories?

42. What is the name of the drug currently under review as a treatment for fear disorders? How does this drug work?

BIOLOGY 21
LAB 5: THE CIRCULATORY SYSTEM

NOTE: THIS LAB *REQUIRES* A PARTNER FOR STATION 3!

Your circulatory system is to your body what BART (or the Light Rail) is to the Bay Area – that is, it is a rapid transport system that carries people to and from work, school, etc. In this analogy, people are molecules such as oxygen, carbon dioxide, ions, macromolecules, etc.; destinations such as work and school are different organs and/or tissues in your body.

We will start our look at the circulatory system by considering blood. Blood is sometimes called the elixir of life (the essence of life). This makes sense when you consider that blood is the medium that carries various types of cells that carry out vital functions, from transporting oxygen to fighting disease.

Blood, to the naked eye, looks little more than a slightly thick red liquid. In reality, blood is much more.

ACTIVITY 1: BLOOD ON THE INTERNET

We will consult a website devoted to hematology that was developed by the Puget Sound Blood Center. Go to Bio 21 on Canvas and under Lab 5, click on "Bloodworks Northwest" (or direct: `www.bloodworksnw.org/medical-services/introduction-to-hematology`)

Scroll down through "Introduction to Hematology" to answer the following:

From "What is blood?"

1. What is blood composed of? _____

From "How blood cells are made"

2. Where is blood made? _____

3. In what 3 bones (or major classes of bones) in your body is most of your blood made?

_____

4. Master cells are also known as what? _____

5. What is unusual about stem cells? (Hint: what have you heard in the news about stem cells

and their potential to treat debilitating conditions such as Parkinson's Disease?). _____

_____

_____

Human Biology: Lab 5

From "Plasma"

6. List the different molecules, etc. that are carried by plasma: _____
_____
_____

From "Red Blood Cells"

7. What is the major function of red blood cells? _____

8. What is hemoglobin? _____
_____

From "White Blood Cells"

9. What is the function of white blood cells? _____
_____
_____

10. What are the five major categories of white blood cells?

a. _____

b. _____

c. _____

d. _____

e. _____

From "Platelets"

11. What is the function of platelets? _____

From nowhere in particular (Google if necessary!)

12. By the way, what is hematology? _____
_____

STATION 1: BLOOD UNDER THE MICROSCOPE

Look at the slide of blood under the microscope to identify the following:
- *red dots:* erythrocytes (red blood cells)
- *purple specs:* platelets
- *bright purple blobs:* leukocytes of various kinds (white blood cells).

13. Which type of blood cell is most common or abundant?_____

14. Which type of blood cell is least common or abundant?_____

15. Which type of blood cell is largest in size? _____

16. Which type of blood cell is smallest in size?_____

Diagram two different types of white blood cells you see under the microscope. Using the chart at the station, identify what kind of white blood cell you've drawn. Also note that some types of white blood cells are far more common than others, so chances are you are looking at one of the more common ones (though you never know...).

White blood cell #1
17a. What type of leukocyte is this?

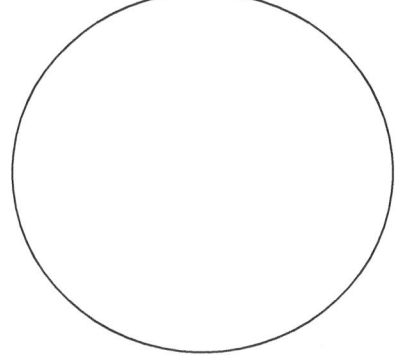

White blood cell #2
17b. What type of leukocyte is this?

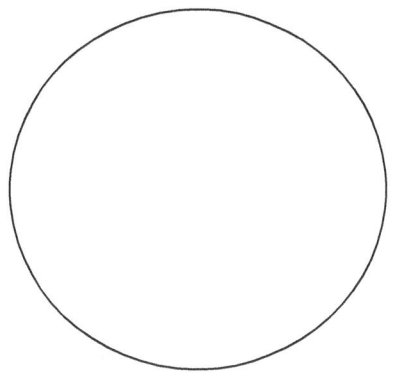

18. What type of white blood cells are most common?_____

STATION 2: DISEASES OF THE BLOOD: ACUTE MYELOBLASTIC LEUKEMIA (AML)

Acute myeloblastic leukemia (AML) is a malignancy of blood-forming tissues that is characterized by the proliferation of <u>immature</u> white blood cells called granular leukocytes. AML may occur at any age, but it primarily occurs in adults and in children below 1 year of age. In AML, malignant cells lose their ability to mature and specialize. These cells multiply rapidly and replace the normal cells. The person becomes susceptible to bleeding and infection as the blood cells lose their ability to function. (Fortunately, new advances in treatment for AML have drastically increased the survival rate for those affected by AML.)

19. There are two things about the blood cells in this slide that are unusual. What are they?

_____

Human Biology: Lab 5

ACTIVITY 2: A DAY IN THE LIFE OF A BLOOD CELL

Blood cells travel a well-worn path in your body. This path occurs as two circuits: one involving your heart and lungs, and the other involving the heart (again) and all the other parts of your body except for the lungs. It's important to remember that these circuits are connected, and the path of a single blood cell is continuous between the two. Watch the following short video at `www.youtube.com/watch?v=NDk8fmIl9V8` (also on the Bio 21 Canvas site under Lab 5) and answer the following:

20. The cardiovascular system is divided into what two parts (or systems)? What is the <u>general</u> pathway of each?

    a. _____

    Pathway _____

    _____

    b. _____

    _____

    Pathway _____

    _____

21. Oxygen-poor blood is found in what two chambers of the heart?

    a. _____

    b. _____

22. Oxygen-rich blood is found in what two chambers of the heart?

    a. _____

    b. _____

23. As a rule, arteries carry oxygen-rich blood, while veins carry oxygen-poor blood. What is

    the exception to this rule? (Hint: listen very carefully to the video clip and pay attention to

    what kind of vessel is carrying what kind of blood.) _____

    _____

    _____

5.4   Human Biology: Lab 5

ACTIVITY 3: A CLOSER LOOK AT THE HEART

24. Now that you're more familiar with the pathway of blood, let's take a closer look at the heart itself. Referring to diagrams in Chapter 13 of your textbook, label the following structures in the diagram below: superior and inferior vena cava, right and left atrium, left and right pulmonary veins, right and left ventricle, pulmonary trunk, left and right pulmonary arteries, and the aorta.

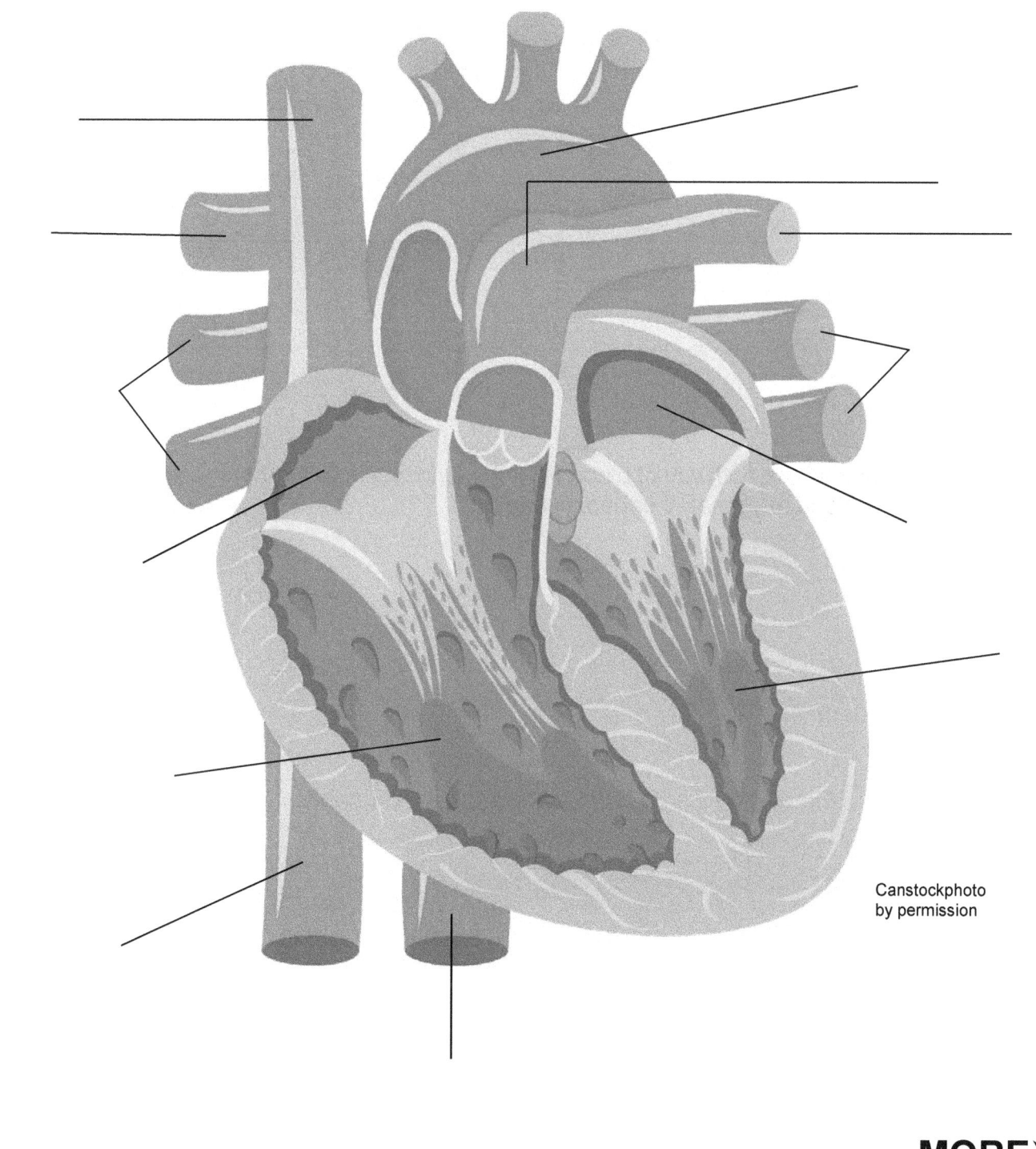

Canstockphoto by permission

MORE▶

ACTIVITY 3: A CLOSER LOOK AT THE HEART (CONTINUED)

25. Next, diagram the pathway of blood as it returns from the body into the right atrium until it exits to the lungs. Is blood oxygenated or deoxygenated? (You can diagram the flow using arrows, for example Structure A ➔ Structure B, etc.) Useful sources of information include Figure 13.1 in your textbook, as well as the video you watched in Activity 2 (www.youtube.com/watch?v=NDk8fmI19V8). You might also find this video useful, too: www.youtube.com/watch?v=ABTvNR59K5Q (also on the Bio 21 Canvas site under Lab 5).

26. Finally, diagram the pathway of blood from the lungs back to the heart and out to the rest of the body. Is blood oxygenated or deoxygenated?

STATION 3: BLOOD PRESSURE & PULSE

The heart pumps blood into a network of vessels (i.e., arteries, veins and capillaries). Your blood pressure is the pressure that blood exerts against vessel walls. When you measure your blood pressure, you are actually obtaining two different kinds of information.

Systolic Pressure. The left side of the heart (in particular, the left ventricle) contracts and sends blood into the aorta. At this point, aortic pressure is at its peak. The systolic reading records this peak pressure, that is, the maximum pressure exerted on the aorta during this contraction. Systolic pressure is the number in the numerator (first number) of the blood pressure reading, e.g., the 120 in "120 over 80" (or whatever your reading is).

Diastolic Pressure. After the blood is pushed out of the heart and into the aorta, the heart relaxes. Diastolic pressure measures the lowest blood pressure in the aorta. Diastolic pressure is the number in the denominator (second number), e.g., the 80 in "120 over 80".

Arteries throughout your body expand and contract as a result of blood moving through your body under pressure. These surges of blood create a pulse. Your pulse is typically easiest to detect at the wrist or neck (but in fact, your pulse is everywhere arteries are!)

You will use the Omron Portable Wrist Blood Pressure Monitor to record your blood pressure and pulse three times each.

a. Place the cuff on your LEFT wrist. Hold your left arm in front of you with your palm facing up. Apply the cuff so that that the monitor is on the inside of your wrist with the display in clear view. The cuff's edge should be approximately ¼ to ½ inch from below your palm (refer to diagram at station). Wrap the cuff comfortably around your wrist. The top and bottom edges of the cuff should be tightened evenly around your wrist.

b. Sit down on the chair at the counter and place your left elbow on the counter so that your wrist with the cuff is at the same level as your heart.

Important: remain still and don't talk while taking your blood pressure!

c. Press the START/STOP button. The cuff automatically inflates to approximately 180 mmHg (about 50 mmHg higher than the average systolic blood pressure reading; if your blood pressure is higher than average, the monitor will momentarily deflate and then re-inflate to 230 mmHg – do not press any buttons while this is happening. It is possible to override the automatic detection and inflate higher than 230 mmHg; however, in no case should you inflate the cuff higher than 280 mmHg).

d. When the cuff reaches your correct inflation level, it begins to deflate automatically. As the cuff deflates, decreasing numbers appear on the display. The Heart Symbol flashes at the onset of oscillation signals.

e. When the Heart Symbol stops flashing, your blood pressure and pulse will display alternately. Record your blood pressure and pulse in the box below (under Trial 1).

f. You will take your blood pressure and pulse two more times in order to obtain an average blood pressure and pulse reading. It is important to wait two to three minutes in between readings (this will allow engorged blood vessels to return to normal). Best way to do this is to trade off using the cuff with your lab partner.

	Blood Pressure	Pulse
Trial 1		
Trial 2		
Trial 3		
Average		

g. On the clipboard at the end of this station, record your average pulse and your age in the age category that applies to you. We'll now consider how a person's pulse is affected by his or her age.

h. Using data from the clipboard, record in the table below the pulse and age for 4 additional people in Age Group III. (We've supplied data below for categories I, II, and IV). If you are over 65, substitute your data for one of the subjects in Age Group IV; write down pulse and age data for 5 students in Age Group III.

		Age	Pulse
Age Group I (1-7)	1	1	117
	2	3	100
	3	2	115
	4	1	120
	5	2	110
Age Group II (8-14)	6	8	100
	7	12	87
	8	9	95
	9	13	86
	10	10	90

		Age	Pulse
Age Group III (15-64)	11		
	12		
	13		
	14		
	15		
Age Group IV (65+)	16	66	70
	17	69	71
	18	70	72
	19	65	76
	20	72	70

ACTIVITY 4: ANALYZING PULSE AND AGE DATA USING EXCEL

Microsoft Excel is on all computers in the Biology 21 lab. Click on the Excel icon on the desktop to open up a new spreadsheet. If you're unfamiliar with Excel, the most important thing to notice is that the spreadsheet appears as a series of "cells," e.g., the cell in the uppermost left corner is cell A1 (i.e., row 1, column A). Directly beneath A1 is A2, and directly to the right of A1 is B1, and so on. When entering data, you enter one number in each cell.

Enter the data from Station 3 for all four age groups in the Excel spreadsheet. First, type the title AGE in cell A1. Then type the title PULSE in cell B1. Next, enter all numbers for "AGE" in column A, and all numbers for "PULSE" in column B (age and pulse for each person must paired; e.g., if you enter John's age in cell A2, make sure you put his pulse in cell B2, etc.).

We'll now plot this information as a scatter plot. A scatter plot is a way of looking at the values of PAIRS of numbers (e.g., age and pulse) in order to visually determine if there is an association between the two (for example, does pulse change with age?).

a. IMPORTANT! <u>First select the cells you want to plot, including the column titles!</u> Position your mouse on cell A1, left click and drag the mouse down to cell B21.

b. With the cells selected, click on INSERT tab (see arrow). You should then see Charts (see circle). From Charts, select the icon representing a scatterplot (below, this is on the bottom right of Charts).

c. In the Scatter window, select the icon on the top left.

d. Type a title where it says Chart Title. Now give your *axes* titles. Click on the **Add Chart Elements** tab (upper left corner) **> Axis Titles > Primary Horizontal**. Double click on the Axis Title at the bottom of your graph and type a title (Age, for example). To add a vertical axis title, click on **Add Chart Elements > Axis Titles > Primary Vertical**. Double click on the Axis Title on the side of your graph and type a title (Pulse, for example).

27. Looking at your graph, describe the trend in your data. _____

_____

28. Why do you think this happens? _____

_____

Click on your graph and **Save as Adobe PDF** to the desktop (it is important to click on the graph first to get just the graph itself to save). Upload the pdf to Canvas under Assignments > Lab 5 > Analyzing Pulse and Age Dating Using Excel. Delete your Excel and PDF file on the Bio 21 computer. **NOTE: The graph is part of the lab – if you do not upload the graph, the lab is considered incomplete and you will receive a 0 for the full lab.**

ACTIVITY 5: THE HEART ON THE INTERNET

Go to Bio 21 on Canvas and under Lab 5, click on the link for the American Heart Association (or go direct: www.heart.org). Search the site for answers to the following:

29. What is atherosclerosis? What are the dangers of atherosclerosis?

30. What are the warning signs of a heart attack? (There are 5 listed.)

31. What are some of the lifestyle changes one could make to prevent heart disease?

32. Revisit the last two questions in Activity 4 (why does pulse rate change with age?). After reading about heart disease (and in particular, arteriosclerosis), how would you now answer this question?

ACTIVITY 6: CHOCOLATE HEARTS: YUMMY & GOOD MEDICINE?

Refer to the article on Canvas under Blood & Circulation/Lab 5 titled "Chocolate Hearts: Yummy & Good Medicine?". Read the article and answer the following questions:

33. According to the article, chocolate contains what kind of antioxidant? _____

34. What two effects do chocolate's antioxidants have on our cardiovascular systems? _____

_____

_____

_____

35. Of the different types of chocolate, which type has the most antioxidant? _____

Please help yourself to some chocolate in the lab (this is science, after all...).

BIOLOGY 21
LAB 6: RESPIRATION & GAS EXCHANGE

NOTE: THIS LAB *REQUIRES* A PARTNER FOR STATIONS 3 AND 4 (WE RECOMMEND HAVING A PARTNER FOR STATIONS 1 AND 2)

We breathe in order to take in oxygen (O_2) and get rid of the waste products produced by our cells, namely carbon dioxide (CO_2). The general term for this is respiration. Taking in air is called inspiration, while blowing air out is called expiration. A respiratory cycle consists of one inspiration and one expiration. The rate at which your body performs a respiratory cycle depends on the levels of oxygen and carbon dioxide in your blood.

Have you ever hyperventilated? That is, found yourself breathing hard but feeling like you still can't get enough "air"? What's actually happening is that your intake of oxygen and output of carbon dioxide are out of balance, and you are exhaling too much carbon dioxide. As a result, your arms, legs and mouth may tingle and feel numb. Hyperventilating is usually not serious, but it is uncomfortable. As you go through this lab, think about why it is sometimes helpful to blow repeatedly into a paper bag when you hyperventilate.

STATION 1: DIFFUSION (Modified from *Biology in the Laboratory*, Helms et al., Freeman & Co., New York).

Small molecules such as O_2, and CO_2 go back and forth across the cell membrane as a result of a process called <u>diffusion</u>.

Diffusion is a simple concept that can be demonstrated by thinking about what happens when you spray yourself with perfume or scented deodorant. If you stand in the same place where you sprayed yourself, eventually the scent disappears. What's going on? Some of the "scent molecules" (technically known as a solute) have landed on you and have been absorbed, while the rest linger in the air – first around you, and then eventually throughout the room.

Here's the rule: molecules move from an area of high concentration (where you sprayed) to an area of low concentration (away from you). We often talk about a <u>concentration gradient</u> as a range of concentration of molecules or "solutes" – that is, a range from high concentration to low concentration. It's pretty easy to think about diffusion in air – since that's the medium we, as air breathing creatures, most readily identify with. However, the same concept of diffusion applies to the movement of solutes in water.

<u>Diffusion is the movement of a solute from an area of high concentration to an area of low concentration.</u>

What to do:
<u>Part I</u>

a. Put the bottom (i.e., smaller) half of a clear plastic Petri dish on the photocopy of a ruler (see sample on lab bench).

b. Measure 30 ml of deionized water (from the left-hand faucet at the sink) into a graduated cylinder. Pour water into Petri dish and let the Petri dish rest for a minute.

c. Add one drop of dye to the center of the Petri dish. Cover the dish with its lid, taking care not to disturb the water (and dye). Start the timer for two minutes.

d. You'll now observe, at 2 minute intervals, how fast the dye diffuses in the water over 10 minutes total. You've already started the timer in step (d); when the timer goes off, record the diameter of the dye spot. Reset the timer for another two minutes and repeat the measurement, for a total of 5 recordings.

Data Table 1

Time in minutes	Diameter (D) in millimeters (mm)
0	0
2	
4	
6	
8	
10	

e. Please dispose of the water and dye down the sink. Rinse and dry the Petri dish, and return it to the station for the next student. <u>Wipe up any spills</u>. Thank you!

f. Plot your data from Table 1 on the graph below:

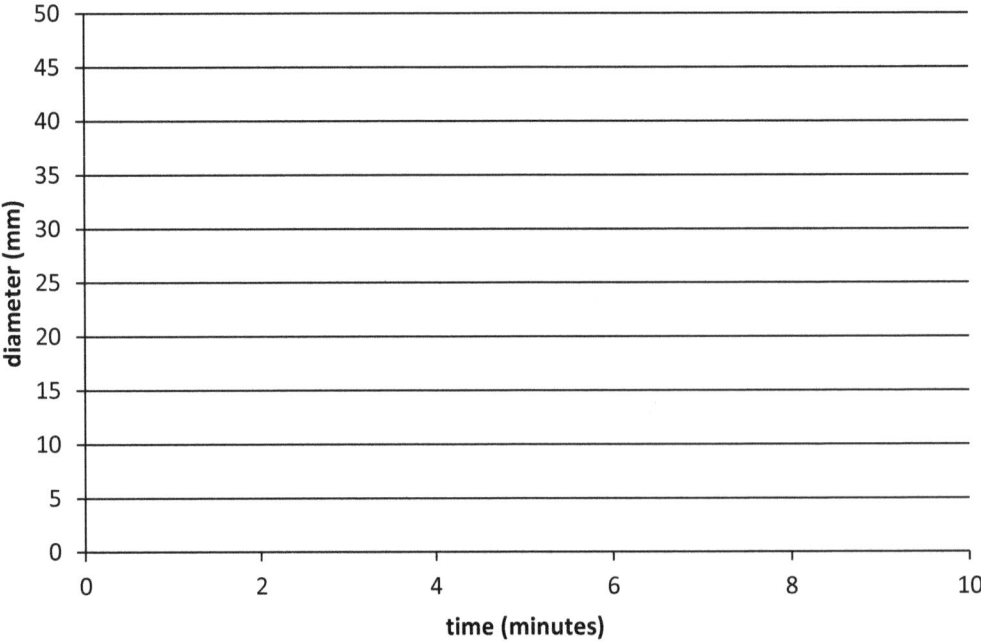

g. Choose three 2-minute intervals and calculate the rate of diffusion from the data in your graph. Use the following equation and record your results in the table below.

$$\text{Rate of diffusion} = \frac{\text{(diameter at end of a 2-minute interval)} - \text{(diameter at start of a 2-minute interval)}}{2 \text{ min}}$$

Time Interval	Rate of Diffusion (mm/min)
Minute _____ to _____	_____
Minute _____ to _____	_____
Minute _____ to _____	_____

1. Does the net movement of molecules slow down as equilibrium is reached? _____

Why? _____

2. Does diffusion eventually come to an end? _____

Why? _____

Part II

Next you will consider what effect temperature has on the rate of diffusion. Before you begin, formulate a hypothesis

What is a hypothesis? A hypothesis is a tentative answer or explanation to a question or problem. Good hypotheses focus on a specific question or problem.

First we identify the variable we are studying and then we create two statements that represent two possible outcomes:

- The variable being manipulated has an effect on the outcome. A prediction of the variable's effect on the outcome becomes your hypothesis. So a hypothesis is a tentative statement that proposes a possible explanation.

- The variable being manipulated has no effect on the outcome, i.e., there is no relationship. This is your NULL hypothesis.

3. In Part II, what two variables are we studying?

a. _____ b. _____

4. An independent variable is something the experimenter decides to vary (manipulate).

Which variable is the independent variable? _____

Human Biology: Lab 6

5. A dependent variable is something that is affected by or responds to the manipulation of an independent variable. Which variable is the dependent variable? _____

Once you've identified your independent and dependent variables, you can construct a hypothesis and null hypothesis.

6. Hypothesis: _____

_____

_____

7. Null Hypothesis: _____

_____

_____

Check your answers to questions 3 – 7 with the lab instructor. Have him or her initial in the box to the right before proceeding to the experiment.

Once you have formulated your hypotheses, you will repeat the procedure for placing the dye in water, this time using a separate dish for cold water and one for hot water. Go to Station 2.

STATION 2: DIFFUSION CONTINUED (Modified from *Biology in the Laboratory*, Helms et al., Freeman & Co., New York).

You can conduct both experiments (cold water and hot water) simultaneously (this works best if you have a partner).

a. Place the bottom (i.e., smaller) half of a clear plastic Petri dish on the photocopy of a ruler (see sample on lab bench).

b. Measure 30 ml of COLD deionized water from the beakers in the ice bucket into a graduated cylinder. Pour this quantity into the Petri dish marked **A**. Measure out 30 ml of HOT deionized water into a graduated cylinder; pour this quantity into the Petri dish marked **B**. Allow water to stand undisturbed for a minute.

c. Add one drop of dye to the center of each dish. Cover the dishes with their lids, taking care not to disturb the water (and dye).

d. Find the rate of diffusion of dye in the water by measuring the diameter (D) in millimeters of the dye spot at 30-second intervals over a total of 5 minutes. Record your results on the next page.

PETRI DISH A (COLD WATER)	
Interval	Diameter (D) in millimeters
30 sec.	0
1 min.	
1½ min.	
2 min.	
2½ min	
3 min.	
3½ min.	
4 min.	
4½ min.	

PETRI DISH B (HOT WATER)	
Interval	Diameter (D) in millimeters
30 sec.	0
1 min.	
1½ min.	
2 min.	
2½ min	
3 min.	
3½ min.	
4 min.	
4½ min.	

e. Please dispose of the water and dye down the sink. Rinse and dry the Petri dish, and return it to the station for the next student. Wipe up any spills. Thank you!

f. Determine the rates of diffusion of the colored liquid in both hot and cold water (using same procedure explained in station 1) and record them below.

Time Interval	Rate (mm/min) COLD	Rate (mm/min) HOT
Minute _____ to _____		
Minute _____ to _____		
Minute _____ to _____		

8. Do your results support your hypothesis? _____

_____

9. What do you conclude about the effects of temperature on the rate of diffusion?

_____

_____

Human Biology: Lab 6

ACTIVITY 1: OVERVIEW OF THE RESPIRATORY SYSTEM
In your textbook, refer to Chapter 14 to answer the following questions:

10. Name the two parts of the respiratory system and the organs/structures included in each.

_____

_____

11. The trachea splits into two passageways called the _____

12. What is the last structure of the air passageway in your lungs? _____

13. What are the functions of the larynx? _____

_____

STATION 3: SPIROMETRY (Adopted from *Human Physiology* by S.I. Fox)

Note: do the first part of this activity (using the spirometer) only if you are healthy. Everyone is still responsible for knowing the following information, however. If you have a cold or otherwise choose not to use the spirometer, you will use another student's chart (posted at the station) to complete the activity starting with "How to Interpret Your Chart" on page 6.8. (Pick a chart that is comparable to you, i.e., similar age and same sex.)

Clinicians use spirometers to measure lung capacities. Spirometers are an excellent means of detecting abnormalities in lung capacities for those with pulmonary disorders such as *asthma* and *emphysema*. With the help of spirometers, health care professionals are able to monitor loss of pulmonary function and track changes in respiratory disease. By monitoring a patient's ventilation, decisions can be made regarding treatment of the following:

- **Obstructive Pulmonary Disease**. *Asthma* and *bronchitis* are obstructive diseases. They cause constriction of the bronchioles and an increase in mucus which increase airway resistance. A more serious obstructive disease, called Chronic Obstructive Pulmonary Disease (*COPD*), is actually two diseases, *emphysema* and *chronic bronchitis*. COPDs both obstruct and restrict airways reducing lung capacity.

- **Restrictive Disorders**. Restrictive disorders cause a reduction in lung capacity. *Tuberculosis* is a common restrictive disorder and is caused by a bacterial infection.

Your total lung capacity has four components:

1. **Tidal Volume (TV)**. This is the volume of air inhaled and exhaled during a normal breath. When you're at rest, tidal volume is at its minimum; during strenuous exercise, tidal volume is at its maximum. Tidal volume is about a pint of "air".

2. **Inspiratory Reserve Volume (IRV)**. This is the volume of air you can voluntarily inhale (again) after inhaling your tidal volume.

3. **Expiratory Reserve Volume (ERV)**. This is the volume of air you can voluntarily exhale after exhaling your tidal volume.

4. **Residual Volume (RV)**. This is the volume of air that cannot be exhaled from the lungs. There is always some air in the lungs, even after you have forcefully exhaled.

You will also see reference to "vital capacity" (VC). Add up TV, IRV, and ERV, and you have your vital capacity (VC).

What to Do: We will measure two of the four lung capacity components: tidal volume and vital capacity. Important: this activity takes coordination because you're recording data over a relatively short amount of time. Read through steps b-e at least once. Go through the motions of steps d, e, f, and g without the drum moving so that you will know what to expect.

In the directions below, we refer to a "*breather*" and the "*facilitator*" for the sake of clarity.

a. Ask the lab TA to load the paper and the drum. Please do not attempt this yourself, as it is very easy to break the drum's motor on the spirometer.

Once the paper is loaded, place the uncapped pen so that it will start writing in the middle of the chart (you need to give the pen room to record your heavy breathing...).

b. *Breather*: place a folded Kim-wipe tissue over your nose and secure the nose clip.

c. *Breather*: using a fresh, disposable mouthpiece, place the mouthpiece in your mouth. You will now be breathing through the apparatus (it feels like you're breathing through a snorkel).

d. *Facilitator*: lift the bellows of the spirometer up about 6 inches and hold until the "breather" starts breathing through the tube (step **e**).

e. *Breather*: breathe in and out normally a few times to get used to the apparatus.

f. *Facilitator*: when the breather is ready, check to see that the switch is on the SLOW position (32 mm/min), and turn the spirometer ON. The breather should breathe in and out for about half a minute. This will record his or her tidal volume.

With the drum still turning, you will now perform the test for vital capacity.
g. *Breather*: at the end of a normal exhalation, inhale as much as possible and then exhale completely to the fullest extent

h. *Facilitator*: stop the recording and first remove the drum from the spirometer and then remove the chart from the drum.

i. *Breather*: remove the mouthpiece and place it in the big red biohazard bin on the floor.

Reverse roles and repeat steps **a-i**.

Place used mouth pieces in the biohazard bag. Thank you!

How to Interpret Your Chart

First notice the numbers on the vertical axis of your chart – these represent liters of air.

Your tidal volume is indicated by the small waves (see cartoon below). To determine tidal volume (before correction), subtract "peak" from "valley," that is, the number of liters at the peak of one wave from the number of liters at the bottom of the next valley. Pick any peak/valley pair that seems representative of your "profile." Write this number below:

14. Tidal Volume Before Correction: _____

We're not done yet, however. The temperature and pressure of the spirometer are different than the temperature of your body and the pressure in your lungs. To correct for this, we need to multiply by a correction factor called BTPS (body temperature, atmospheric pressure, saturated with water vapor). When the room is at "normal" temperature (about 68°F), BTPS is 1.1. Multiply "Tidal Volume Before Correction" by 1.1 to get tidal volume.

15. Tidal Volume After Correction: _____

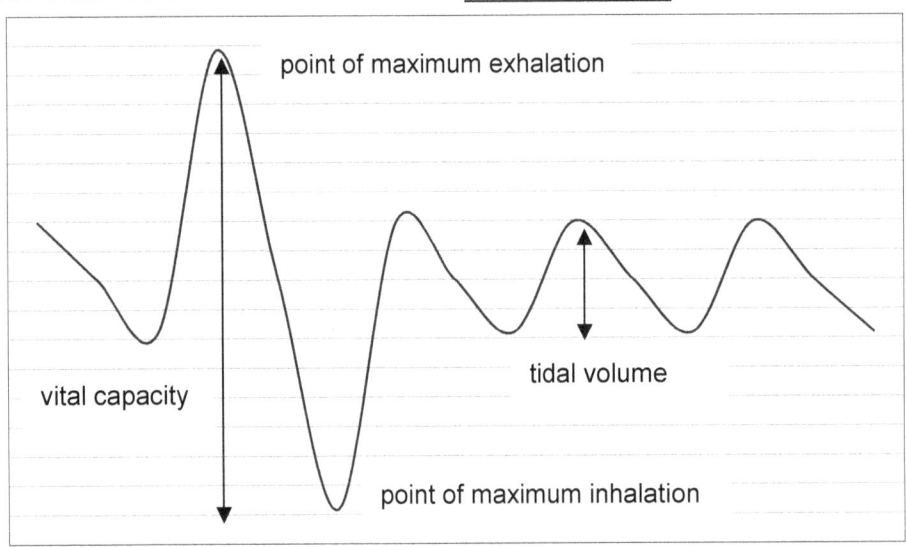

Next, mark the highest peak on your chart "point of maximum exhalation" and the lowest valley "point of maximum inhalation."

To determine vital capacity (before correction), subtract liters at "point of maximum exhalation" from liters at "point of maximum inhalation."

16. Vital Capacity Before Correction: _____

Now multiply "Vital Capacity Before Correction" by the BTPS factor (1.1) to get vital capacity.

17. Vital Capacity After Correction: _____

Evaluating Your Data

Vital capacity differs from person to person, depending on age, sex, and size. Usually:

- ❖ Taller people have higher vital capacities than shorter people.

- ❖ Men have higher vital capacities than women (and not just because men tend to be taller than women).

- ❖ Younger people have higher vital capacities than older people.

In order to evaluate your vital capacity, you will find laminated Vital Capacity Charts on the lab bench near the spirometers. First, enter the following data for yourself (or for the person whose data you are using):

18. Sex: _____

19. Age: _____

20. Height: _____ in inches.

Using this information, refer to the charts at the spirometry station (Station 3) to locate the predicted vital capacity for a person of your sex, age, and height.

21. Your predicted vital capacity: _____

22. Divide your observed vital capacity (after correction) which you recorded in #17 above by your predicted vital capacity (#21), and multiply by 100.

$$\frac{\text{Vital Capacity (after correction)}}{\text{Predicted Vital Capacity}} * 100\% = \underline{\qquad}$$

What does this mean? An observed vital capacity that is *repeatedly* and *consistently* below 80% of the predicted value may indicate a restrictive lung disease.

Something to Think About
Emphysema patients have difficulty expelling air from their lungs because their residual volume is increased. What effect would this have on their:

23. Total lung capacity? _____

24. Vital capacity? _____

25. Effectiveness of gas exchange? _____

STATION 4: CO_2 AND YOU

As you learned in lecture, you breathe in oxygen and exhale CO_2 (carbon dioxide) as a cellular waste product. In this experiment, we'll look at how increased muscle metabolism during exercise results in an increase in CO_2 production.

In this exercise, you will prepare two beakers of solution containing water, a base, and phenolphthalein. Phenolphthalein is used to give us a general idea about the pH of a solution. Phenolphthalein is PINK in alkaline solutions and CLEAR in neutral or acidic solutions. An easy way to measure the carbon dioxide in your breath is to blow bubbles in a solution of water and base containing phenolphthalein. When you do this, the CO_2 from your breath mixes with water (H_2O) to create carbonic acid. Therefore, the solution should turn CLEAR because it has become more acidic as a result of the mixture of your CO_2 and the water.

a. Fill a beaker with the following

- 200 ml of deionized water
- 5.0 ml of 0.10 NaOH (sodium hydroxide, a base)
- 3 drops of phenolphthalein indicator

b. Divide the solution you made in step 1 into two smaller beakers and stir. Put your two beakers into a small tray and put a piece of paper with your names in the tray.

c. While sitting quietly, have one person exhale through a straw into the solution in the FIRST beaker, while the other person times how long it takes to turn the solution from PINK to CLEAR. Record your observation in the table below (first line).

	Time for solution to turn clear
When sitting quietly	
After vigorous exercise	

d. Have the person who exhaled into the beaker now go run (or walk) up and down the stairs for 2 to 5 minutes. Without stopping to rest, go onto step e.

e. Have the "stair climber" now exhale through a straw into the solution in the SECOND beaker, while the other person times how long it takes to turn the solution from PINK to CLEAR. Record your observations in the table above (second line).

f. Dispose of your solutions in the sink. Rinse and dry the beakers for the next students.

Questions

26. Was there a difference in the amount of time it took for the solution to turn CLEAR after exercise?_____

27. Why do you think there was (or was not) a change?

ACTIVITY 2: LUNGS ON THE WEB

Go to Bio 21 on Canvas and under Lab 6, click on American Lung Association, or go directly to www.lung.org. In the search bar, type "lung cancer" and answer the following:

28. What is lung cancer?

29. What causes lung cancer?

30. Who is at higher risk for lung cancer?

31. How is lung cancer detected?

32. How is lung cancer treated?

33. How is lung cancer prevented?

BIOLOGY 21
LAB 7: URINARY & ENDOCRINE SYSTEMS

Note: This lab is completed on-line and at home. <u>You will need to come to lab to show your completed lab assignment to a lab instructor and take the quiz to receive credit.</u> The lab will be open fewer hours this week, so be sure to consult the schedule on Bio 21 on Canvas.

This week we cover the urinary (or renal) system, and then take a broader look at our body's hormones and their targets (including the kidneys).

I. Urinary System

The kidneys play many essential roles: they excrete the nitrogenous waste products, regulate water balance and blood electrolyte concentration, and help to maintain the body's pH within a range that allows normal physiological processes. The kidneys also secrete erythropoietin, a hormone that promotes the production of red blood cells by the bone marrow and renin, an enzyme that affects blood pressure. You'll examine your urine for signs of abnormal kidney function, and other diseases detectable in the urine. Finally, you'll take a closer look at diabetes.

ACTIVITY 1: OVERVIEW OF THE URINARY SYSTEM

Refer to Chapter 17 (The Urinary System) in your textbook to answer the following:

1. The main organs of excretion are the _____

2. Urine is carried from the kidneys to the bladder in what tube? _____

3. Urine is carried from the bladder to the exterior through what tube? _____

4. What is the name of the artery that supplies blood to the kidney? _____

Next, watch the video clip on the Urinary System at `www.youtube.com/watch?v=qxb2_d9ilEw_` (you can also get to this video from Bio 21 on Canvas under Lab 7). Watch the first three minutes of this video.

5. Following digestion, excess nutrients, salts, minerals, water, drugs, and toxins end up in the blood stream. Name two toxins that are cellular waste products:

_____   _____

6. What happens if excess substances, toxins, and cellular wastes build up in the blood?

_____

_____

7. What organ is responsible for maintaining proper chemical balance? _____

Human Biology: Lab 7

ACTIVITY 2: TESTING URINE

Your instructor will distribute a Multistix and a reference card in class or lab. Be sure you have both the Multistix and the reference card before starting this!

You will use a Multistix at home to test your urine for 10 different properties. You will then look up the meaning of the other tests using Internet and text sources. Here is a good video to watch first `www.youtube.com/watch?v=TuWiy4_VDWY` (also under Lab 7 on Canvas).

8. In the video, the narrator indicates that the urine is testing positive for blood. Suppose this sample is from a woman. Can you think of a reason why blood might be in urine? (The sample, in this case, would be completely normal.) _____

READ STEPS 1 & 2 <u>BEFORE</u> STARTING - THIS TEST MUST BE DONE QUICKLY!

<u>Step 1</u>: Dip the Multistix into a paper cup containing your urine and remove it. You want to make sure the Multistix is coated with urine but do not let it soak in the urine. Work quickly, since you need to start reading the Multistix 30 seconds after it comes in contact with urine.

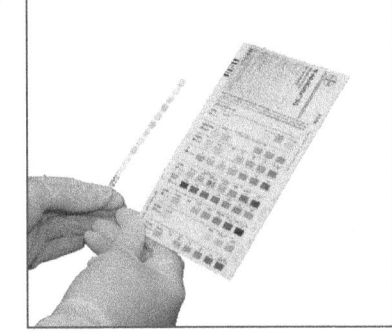

<u>Step 2</u>: Read your urine Multistix. Hold your Multistix so that the blue band (i.e., test for glucose) is on the bottom. Orient the chart so that the glucose test is likewise at the bottom. Simply read up, comparing the color for each test on your Multistix with the chart. Record your results in the table below (you can also circle the color on the card if you like):

TABLE 1: TEST DATA

	READING (e.g., negative or value)	Meaning of your reading (complete AFTER you have finished reading the entire strip (see Activity 3 on next page)
Leukocytes (LEU)		
Nitrite (NIT)		
Urobilinogen (URO)		
Protein (PRO)		
pH		
Blood (BLO)		
Specific Gravity (SG)		
Ketone (KET)		
Bilirubin (BIL)		
Glucose (GLU)		

Once you are done inspecting your urine, dispose of it down the toilet. You can discard (or recycle) your urine reference card. Dispose of your Multistix in the regular trash.

ACTIVITY 3: UNDERSTANDING YOUR MULTISTIX READINGS

9. Complete the column in your data table on the previous page, "Meaning of your reading," by consulting `https://en.wikipedia.org/wiki/Urine_test_strip` (Also on Bio 21 Canvas site under Lab 7.)

ACTIVITY 4: HOMEOSTASIS AND WATER BALANCE

Watch the last 5 minutes of the video "The Virtual Body: Homeostasis" on youtube at `www.youtube.com/watch?v=QKT47A-LBj4` (Also on Canvas under **Lab 2** – you watched the first part of this video in an earlier lab) Fast forward to 12:55. Answer the following:

10. What hormone regulates water content?

11. What two ways does the hypothalamus respond to dehydration?

12. The kidneys' "cleaners" are called _____.

ACTIVITY 5: DIABETES AWARENESS

Go to Bio 21 on Canvas and under Lab 7 click on Diabetes Awareness (or directly at `www.youtube.com/watch?annotation_id=annotation_3720898155&feature=iv&src_vid=jHRfDTqPzj4&v=X9ivR4y03DE`). Answer the following:

13. What happens to a normal individual when he or she eats a carbohydrate such as an apple? The apple is broken down into _____ (hint: this is the simplest form of sugar). The pancreas then secretes _____ (hint: this is a hormone) that allows _____

As you watch the rest of this clip, make notes in the table on the next page to distinguish between Type 1 diabetes and Type 2.

	Type 1 Diabetes	Type 2 Diabetes
Is insulin produced? If it is produced, what is the problem?		
Which type is more common? What age and ethnicity groups are most affected by each?		
What happens to glucose in the blood stream with this type of diabetes?		
What are the main symptoms?		
When do these symptoms appear?		

II. Endocrine System

The endocrine system is made up of several glands and organs found throughout the body that secrete hormones into the blood. The body uses these hormones to control many different bodily functions including (but not limited to): respiration, metabolism, reproduction, sensory reception, movement, sexual development, and growth.

Here you will review the concept behind negative and positive feedback systems, take a look at some of the organs and hormones comprising the endocrine system, and finally watch an interesting TedX talk on obesity, diabetes and insulin resistance.

ACTIVITY 6: NEGATIVE AND POSITIVE FEEDBACK LOOPS
Go to Bio 21 on Canvas, and under Lab 7 click on "Feedback Loops" or go directly to www.youtube.com/watch?v=CLv3SkF_Eag

14. A _____ loop brings your body closer to a target set point, whereas, a _____ loop amplifies a response moving the body away from a target set point.

15. What is homeostasis? _____

_____

_____

16. List the three homeostatic conditions in the human body that are maintained through negative feedback loops. _____

_____

17. What is an example of positive feedback in humans? _____

ACTIVITY 7: OVERVIEW OF THE ENDOCRINE SYSTEM

Go to Bio 21 on Canvas and under Lab 7 click on "The Endocrine System" (or go directly to youtube.com/watch?v=-S_vQZDH9hY). Fill out the table below:

Gland	Hormone(s)	Action(s)
Pineal		
Anterior Pituitary		
Posterior Pituitary		
Thyroid		
Parathyroid		

MORE>

Human Biology: Lab 7

Gland	Hormone(s)	Action(s)
Pancreas		
Adrenal cortex		
Adrenal medulla		
Ovary		
Testes		

ACTIVITY 8: SOMETHING TO CONSIDER: "REVERSING TYPE 2 DIABETES STARTS WITH IGNORING THE GUIDELINES" BY SARAH HALLBERG

Go to Bio 21 on Canvas, and under Lab 7 click on "Hallberg TedX" or go directly to `www.youtube.com/watch?v=da1vvigy5tQ`

18. What is the main point made by Dr. Hallberg in this talk?

19. What are Dr. Hallberg's 5 recommendations?

BIOLOGY 21
LAB 8: THE IMMUNE RESPONSE

THIS LAB *REQUIRES* A PARTNER FOR STATION 3!

We've spent several weeks discussing the amazing process of homeostasis in the human body. These systems have been evolving for millions of years, allowing humans (and other species) to adapt and function in different internal and external environments. As you're no doubt aware, we're under constant siege from bacteria, viruses, and other infectious organisms which, were it not for the body's immune system, would disrupt homeostasis in the body, causing disease and certain death. If you're getting over a cold, or holding the latest flu virus at bay, you can thank your body's immune response for helping you out.

In this lab, we also consider limitations of the immune response; namely, what happens when we intentionally introduce foreign tissue to the body, as in blood transfusions or organ transplants. Understanding when and why the body accepts (or rejects) these introductions is also the domain of the immune response.

STATION 1: LEUKOCYTES UNDER THE MICROSCOPE

Recall that there are three types of blood cells: red (erythrocytes), white (leukocytes) and platelets (thrombocytes). Here we will concentrate on leukocytes, cells that play an important role in our body's response to disease. Using the chart on display, identify and draw the following types of leukocytes. See Table 13.1 in your textbook for the functions of these cells.

<u>Eosinophil</u>

1. What is the function of the eosinophil? _____

_____

_____

_____

<u>Neutrophil</u>

2. What is the function of the neutrophil? _____

_____

_____

_____

Lymphocytes

3. What is the function of the lymphocyte? _____

_____

_____

_____

_____

STATION 2: INFECTIOUS DISEASES OF THE BLOOD: MONONUCLEOSIS

Mononucleosis is a viral infection causing high temperature, sore throat, and swollen lymph glands, especially in the neck. It is typically caused by the Epstein-Barr virus (EBV) which is a member of the herpes virus family. Mononucleosis caused by EBV is the most frequently encountered type and is responsible for approximately 85% of infectious mononucleosis cases. The infection is probably transmitted by saliva. While peak incidence occurs in 15- to 17-year-olds, the infection may occur in any age, typically from 10 to 35 years. "Downey Cells" are lymphocytes infected by Epstein Barr virus. <u>With a purple pen or color pencil, color the Downey Cells in this picture.</u>

4. Looking under the microscope, what is unusual about this cell compared to the other

white blood cells you see around it? _____

ACTIVITY 1: BOZEMAN SCIENCE'S "THE IMMUNE SYSTEM" (~14 minutes)
Go to www.youtube.com/watch?v=z3M0vU3Dv8E (also on Bio 21 Canvas site under Lab 8). As you view this short video, answer the following:

Non-specific Immune Responses
5. What are four characteristics of the skin that make it a good "wall" to prevent the invasion of pathogens such as viruses and bacteria?

6. What happens if a virus or bacteria gains entry to your body through, say, a cut?

MORE▶

ACTIVITY 1: BOZEMAN SCIENCE'S "THE IMMUNE SYSTEM" (cont.)

Specific Immune Responses

7. The body fights antigens with _____.

8. What are antibodies? What is important about the different shapes of antibodies?

9. Where are B-lymphocytes made?

10. How do B-lymphocytes produce antibodies?

11. What are T-lymphocytes? How do they work?

There's a lot of detail from ~9 minutes to the 12:15 minute mark. You do not need to know the specifics of the information presented here.

12. At the ~12:15 minute mark, Mr. Andersen talks about our body's ability to "learn" specific immune responses. Pay attention to the information from this point to the end of the video. Why do you think we haven't been able to develop an effective vaccine for the flu?

STATION 3: BLOOD IS THICKER THAN WATER (AND MORE CLUMPY, TOO). *THIS ACTIVITY REQUIRES AT LEAST ONE PARTNER!*

The body's most powerful and important line of defense against infection is the immune response. The immune system can recognize foreign substances in the body (pathogens) and destroy or inhibit them. This is what keeps bacteria and viruses from overwhelming us.

What allows our immune system to distinguish between body cells (self) and foreign invaders (non-self)? ANTIBODIES!

Antibodies (what our bodies produce to destroy or inhibit invaders) lock onto antigens. As a result, antibodies and antigens clump together. This process is called agglutination. These big clumps are then too big to enter the body's cells. These clumps also attract white blood cells, which engulf and digest the clumps.

This is great if the pathogen is a disease. But what about an "intentional invader"? This is essentially what happens when you receive a blood transfusion. It is essential that the body NOT perceive the blood you're receiving as an invader. Otherwise, you can die. In humans, there are two types of antigens on red blood cells: **Antigen A** and **Antigen B**, and there are four types of blood: **A**, **B**, **AB**, and **O**.

- If you have type A blood, you have Antigen A.
- If you have type B blood, you have Antigen B.
- If you have type AB blood, you have both Antigen A and Antigen B.
- If you have type O blood, you don't have either the A or B blood antigen.

You can also have ANTIBODIES that recognize foreign blood proteins (antigens).

- If you have type A blood, you have anti-B antibodies.
- If you have type B blood, you have anti-A antibodies.
- If you have type AB blood, you have neither anti-A nor anti-B antibodies.
- If you have type O blood, you have both anti-A and anti-B antibodies.

	Group A	Group B	Group AB	Group O
Red blood cell type	A	B	AB	O
Antibodies present	Anti-B	Anti-A	None	Anti-A and Anti-B
Antigens present	A antigen	B antigen	A and B antigens	None

Another way to think about this is that
- ❖ O blood can be given to anyone since it doesn't have any antigens that the receiver's blood would try to attack. <u>O is the universal donor</u>.

- ❖ If you have AB blood, you're in luck! You can receive any kind of human blood since your body doesn't produce anti-A or anti-B antibodies. <u>AB is the universal receiver</u>.

Blood Type	Antigen Present on Red Blood Cells	Antibodies Produced	Donor Can Safely Receive These Types:
A	A	anti-B	A, O
B	B	anti-A	B, O
AB	A, B	None	AB, O, A, B
O	None	anti-A, anti-B	O

In this activity, we'll do an experiment to demonstrate how this works. NOTE: this experiment uses FAKE blood.

Steve's Story...

Steve is brought into the emergency room after walking through a sliding glass door...He lost quite a bit of blood and needs a transfusion. He has type A blood. The emergency room has four different bags of blood from four donors, but unfortunately someone in the lab forgot to label them (oops). You, the tireless technician, need to figure out which of the four bags of blood Steve can safely receive (so he can live long enough to sue the hospital).

Think the Problem Through

13. If Steve has type A blood, what kind of antigens does his blood have? _____

14. What kind of antibodies does his blood have? _____

15. What kind of blood can Steve safely receive? _____

16. What kind of blood will cause Steve's blood to clump (thereby killing him...)? _____

What To Do

In your kit, you will find the following materials for this exercise:
- glass block with 4 Hema-tags
- toothpicks (8)
- fake blood for the following fictitious individuals: Wiley Smith, David Smith, John Smith, and Jane Smith (4 bottles total)
- antiserum (2)
- timer (1)
- pen (1)

a. Label the glass block with the four Hema-tags: one space for Wiley Smith, the second space for David Smith, the third space for John Smith, and the fourth space for Jane Smith.

b. Make sure the caps are on bottles securely and shake each one a few times.

Note: You'll now create a series of different blood/antiserum mixtures to see which ones agglutinate (clump). <u>The key to success of this experiment is using just a single drop of each.</u>

c. Place <u>one</u> drop of anti-A antiserum in all four anti-A circles on the microscope slides.

d. Place <u>one</u> drop of anti-B antiserum in all four anti-B circles on the microscope slides.

e. Add one drop of Wiley Smith's blood to each drop of antiserum on his slide. Mix the blood and antiserum for two minutes, using the wide end of the tooth pick. Use a separate toothpick for each antiserum (otherwise you will contaminate the samples, and Steve could likely end up dead...).

f. After two minutes, check to see if agglutination has occurred in either circle. If agglutination occurred (i.e., you see clumps or the liquid appears stringy), the result is positive; if the blood is clear, the result is negative. A positive result indicates that an antigen is present – use your powers of deductive reasoning (and the information and table on page 8.4 and 8.5) to determine which antigen(s) are present. Fill in the information in the table below.

g. Repeat steps f and g for each of the other three blood samples.

Donor	Reaction to Anti-A antibodies	Reaction to Anti-B antibodies	What Antigens Are Present?	Safe to Transfuse?
Wiley Smith				
David Smith				
John Smith				
Jane Smith				

PLEASE <u>THOROUGHY</u> RINSE THE GLASS BLOCKS AND TOOTHPICKS WITH TAP WATER (DO NOT USE SOAP), DRY THEM, AND RETURN THEM TO THE KIT TRAY.

Analyzing Results

17. What is Wiley Smith's blood type? _____

18. What is David Smith's blood type? _____

19. What is John Smith's blood type? _____

20. What is Jane Smith's blood type? _____

21. Who can safely give Steve blood? _____

MORE ▶

22. What happens if an incompatible blood type is transfused into a patient? What are some medical problems that can arise?

23. If John Smith needs a transfusion, from whom (in the general population) could he safely receive blood? Why?

24. Can any of the donors give blood to all of the other donors? If so, which one and why?

ACTIVITY 2: Your Mother Was Right…

Go to Bio 21 on Canvas and under Lab 8, read the article "Your Mother Was Right" and answer the following questions:

25. According to the article, what is one of the most effective ways of staying healthy?

26. How do pathogens (such as viruses and bacteria) on our hands affect our health? Name a few specific pathways into the body.

27. Name two reasons why antibacterial soaps may do more harm than good. (The article gives one explicit reason; the other, you'll have to read between the lines – hint, read paragraph 5 on the first page carefully).

BIOLOGY 21
LAB 9: REPRODUCTION & STDS

The reproductive system is a vital key in the success of life on earth – if we cannot reproduce ourselves, our species will become extinct. This week, we will study the Reproductive System by first considering the anatomy of both the male and female reproductive systems, and then the hormonal mechanisms behind reproduction.

ACTIVITY 1: OVERVIEW OF THE MALE REPRODUCTIVE SYSTEM

Refer to Section 19.2 in your textbook. Below, diagram the pathway of sperm through the male reproductive tract. Use arrows to show the path, i.e., Structure A → Structure B, etc.

1. _____

_____

_____

STATION 1: QUIZ YOURSELF ON THE MALE REPRODUCTIVE SYSTEM

There are several models showing the anatomy of the male reproductive system. Make sure you understand the function of the following anatomical features (see also information in Section 19.2 of your textbook):

Part	Function
2. Prostate gland	
3. Urethra	
4. Penis	
5. Seminal vesicle	
6. Vas deferens	
7. Bulbourethral gland	
8. Epididymis	

Human Biology: Lab 9

STATION 2: QUIZ YOURSELF ON THE FEMALE REPRODUCTIVE SYSTEM

There are several models showing the anatomy of the female reproductive system. Identify the function of the following anatomical features (see Section 19.3 in textbook):

Part	Function
9. Ovary	
10. Clitoris	
11. Vagina	
12. Cervix	
13. Uterus	

ACTIVITY 2: THE MENSTRUAL CYCLE

View a short video at www.youtube.com/watch?v=1_wX285vrrU (also on the Bio 21 Canvas site under Lab 9) and answer the following:

14. What are the three main "players" in the female reproductive cycle? _____

_____

15. At the beginning of a new menstrual cycle, the nerve cells in the hypothalamus secrete _____ which then causes the pituitary to secrete _____

16. FSH then travels to the ovary where it stimulates the formation and growth of the ovarian follicle. The follicle consists of an _____ and other cells that secrete _____.

17. As the follicle matures, the hypothalamus secretes more GnRH which stimulates the pituitary to secrete what hormone? _____. When this hormone peaks mid-cycle, what happens? _____

18. The increase in what hormone mid-cycle triggers ovulation? _____

19. What does the hormone progesterone do? _____

20. What happens to the uterus lining if the egg is not fertilized? _____

STATION 3: BREAST & TESTICULAR CANCER DETECTION

There are many different types of cancers of the reproductive system. While some of these cannot be detected through self-examinations (ovarian, cervical and prostate, to name a few), others can. Self-examinations are possible, and recommended, for the detection of breast and testicular cancer. This station was designed for you to practice correct examination techniques for the breast and testicles. **Important!** It's not enough to know how to do a self-examination on yourself! Learn how to feel for masses in both testicles and breasts. You could save your partner's life by knowing this!

A. Breast Examination

Follow the directions on the poster to detect the lumps in the breast model. Use the outline on the right to indicate the location of the lumps.

Using the information on the poster, answer the following.

21. When is the best time of the month to do a breast examination?

22. Why should the armpit and collar bone be examined in addition to the breast?

B. Testicular Examination

Follow the directions on the poster to detect the lumps in the testicle model. Use the outline on the right to indicate the location of the lumps.

Using the information in the poster, answer the following.

23. Why should you do a testicular self-examination in the shower or bath?

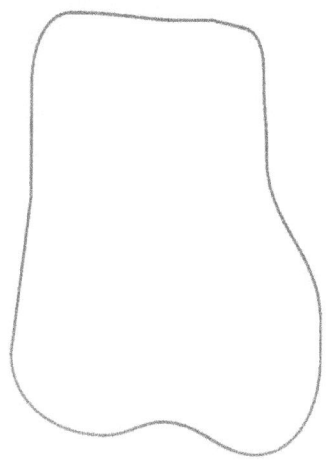

24. Complete the following statement: "You should feel for any small lump about the testicle, particularly those about the size of a _____, on the _____ or _____ of the testicle."

"Google" the internet for the answer to the next question (if you don't know it already):

25. What famous cyclist successfully fought advanced testicular cancer and went on to win 7 consecutive Tour de France titles (until they were taken away, that is)? _____

Human Biology: Lab 9

STATION 4: USING CONDOMS (originally from www.healthsquare.com which is sadly now defunct. Check out www.youtube.com/watch?v=_8TJ6gvbSvo&has_verified=1).

Condoms are essential in preventing the spread of sexually transmitted diseases. A good brand of condoms, used properly, has a 97% effectiveness rate against pregnancy and a 98 to 99% effectiveness rate against HIV. Everyone – men and women – should know how to use a condom. So here's your opportunity to learn how, or perhaps to hone your skills...

Tear open one of the condom envelopes. First notice how it's rolled up. The condom is placed over the tip of an erect penis (one that's hopefully more erect than the model...) and then rolled down the shaft of the penis. Always handle the condom carefully so you don't tear or puncture it – these aren't party balloons. Leave about 1/2 inch of space at the tip of the condom. This leaves room to catch semen and sperm so the condom doesn't break. It's much safer to put the condom on loosely at the top of the penis to allow for this space than to pull and stretch the condom once it's on. Condoms can and will break, making them useless.

VERY IMPORTANT: if you (or your partner) put the wrong side of the condom on the penis so that it won't roll down, throw the condom out and start over. Why? Whatever is on the penis (whether it's virus, bacteria, or a small amount of sperm from pre-ejaculate) will transfer to the condom. If the condom is then flipped over and rolled down, what was on the inside is now on the outside – and guess what? There goes your protection.

Putting the condom on is about as far as we can go with this activity in the Bio 21 lab...Please put the unraveled condoms and wrappers in the trash. Once you've done that, keep reading.

Men: after sex, the condom should stay on you as you pull out of your partner. Do not wait until the penis is flaccid (soft) again, or otherwise the condom will leak or slip off. Use a condom only once. Then throw it away. Use a new condom every time you have sex.

If you need to wet the condom, use a lubricant with a water base (such as K-Y). Never use lubricants such as petroleum jelly, cooking oil, lotion, etc. These make the latex in the condom permeable, and can also cause irritation in your partner.

Many condoms come already coated with spermicidal lubricant. Some condoms are made of an animal membrane instead of latex. These condoms will help prevent pregnancy but do not protect against STDs such as HIV infection.

Store condoms in a cool, dry place. Heat will weaken the latex and can cause the condom to break. If you keep condoms in your pocket, purse, wallet, car, etc., make sure they're fresh. Always carry at least two – in case you put it on wrong side out (see above) or it breaks. Don't use condoms that are old, cracked, sticky, brittle, or discolored. Likewise, don't buy cheap condoms. Always look at the expiration date on the box!

ACTIVITY 3: STDs ON THE INTERNET

Go to www.cdc.gov/std to research STDs and answer the following (see next page).

MORE>

	Bacteria or virus?	Symptoms in men	Symptoms in women	Curable or treatable? Be specific – give name(s) of drugs.	Long term risks of disease.
Chlamydia					
Gonorrhea					
HIV/AIDs					

	Bacteria or virus?	Symptoms in men	Symptoms in women	Curable or treatable? Be specific – give name(s) of drugs.	Long term risks of disease.
Herpes					
Hepatitis B					

26. All STDs can be prevented. You've all heard about "safe sex" (or more aptly, "safer sex"). List THREE ways safer sex can be practiced.

a. _____

b. _____

c. _____

ACTIVITY 4: BIOLOGY OF YOU IN-DEPTH: HPV Vaccine on WebMD

In 2005, the discovery and subsequent implementation of a vaccine to prevent HPV was heralded by the National Cancer Institute as a "major public health success story." Go to www.WebMD.com and type in 'HPV vaccine'. Go to **HPV/Genital Warts**, then **Overview** (note: there are two pages under Overview).

27. While the HPV infection can go away on its own, some types of HPV can lead to cancer.

What kinds of cancer have been linked to HPV? _____

28. Of the sexually transmitted strains of HPV, what two are particularly high risk and account for 70% of cervical cancers? _____

29. Why doesn't a condom fully protect someone from the virus? _____
_____

30. What three vaccines are available to prevent some strains of HPV? _____
_____

31. Go back to **HPV/Genital Warts** and click on **"HPV in Men"**. Why should boys and young men be vaccinated against HPV?_____
_____
_____

32. Finally, go back to **HPV/Genital Warts** and click on **Symptoms & Tests** and/or **Treatment**. How is HPV diagnosed? _____
_____
_____
_____

Human Biology: Lab 9

BIOLOGY 21
LAB 10: DEVELOPMENT & MITOSIS

Note: Lab this week is completed on-line. You will need to come to lab to show your completed lab assignment to a lab instructor and take the quiz to receive credit. The lab will be open fewer hours this week, so consult the schedule on Bio 21's Canvas site or in the lab.

ACTIVITY 1: MITOSIS ON THE WEB

We'll use the University of Arizona's Biology Project website to learn about mitosis, or cell division (you can also refer to Figure 21.3 in your textbook). Go to Bio 21 on Canvas and under Lab 10 click on "Mitosis on the Web" or go directly to
`www.biology.arizona.edu/cell_bio/tutorials/cell_cycle/main.html`

Although DNA was covered earlier this semester, the Biology Project has information on DNA that will be a helpful review for you and which will also help your understanding what's happening in mitosis. Therefore, do the following activities on the mitosis page:

- DNA Basics
- The Cell Cycle
- Mitosis

Run any animations you see links to – a picture really is worth a thousand words in this case. The self-test at the end is a bit advanced, but if you're adventurous, go for it.
Things you should know about mitosis include:

1. What is the purpose of mitosis?

2. What are the stages of mitosis (i.e., what is happening in each stage)?

3. What is the final result of mitosis?

Human Biology: Lab 10

ACTIVITY 2: THE EMBRYO DURING EARLY DEVELOPMENT

Go to Bio 21 on Canvas, and under Lab 10 click on "The Multi-dimensional Human Embryo" (or go directly to embryo.soad.umich.edu). (If you get a "page not available message", click "Try Again" and the page should load.)

4. What are "Carnegie stages"? (Scroll down to the bottom of the main page to see this.)

5. What is meant by the term "postovulatory age"?

Before we look at specific stages, view a time lapse animation of stages 13 - 23 for an idea of what occurs during these stages in development. Go to Bio 21 on Canvas, and under Lab 10 click on Time Lapse or go directly to embryo.soad.umich.edu/resources/morph.html

Stage 13 Embryo

Go back to embryo.soad.umich.edu. Click on any of the embryo images in the picture strip, and then click on "13" (for Carnegie Stage 13).

6. How big is the human embryo at Stage 13? _____ How old in days? _____

Click on a few of the images of the "Stage 13" embryo to see what it looks like at this stage.

Stage 16 Embryo

Go back to the main page and click on the "16" embryo (for Stage 16) and read its text. Click on the first photograph to examine features at this stage. Then click on the first image under "Animations" to view the embryo in 3-D (the pause button may be useful).

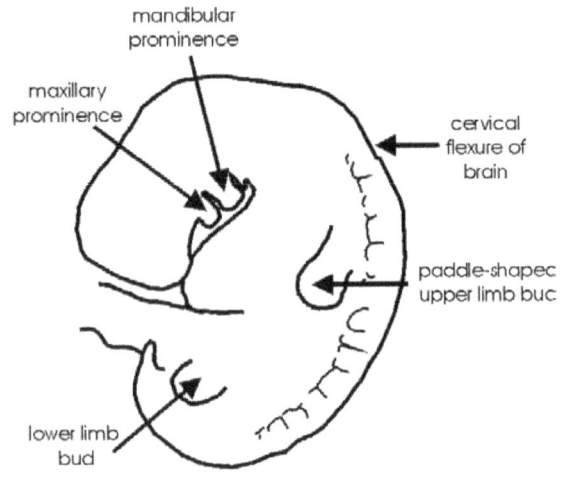

The embryo at this stage exhibits at least 2 sets of obvious "pharyngeal arches" that will become parts of the head and neck later on, a prominent heart "bulge", and 2 sets of limb buds by this time. The diagram on the right may help you identify some of the structures in the Stage 16 embryo.

7. At Stage 16, is the tail-like appendage visible at this stage of development?_____

8. How might this support or refute the idea that complex animals (like humans) gradually evolved from more primitive ones?

ACTIVITY 2: CHANGES IN THE EMBRYO DURING EARLY DEVELOPMENT (continued)

Stage 23 Embryo

Return to the main page and click on the "23" (for stage 23 embryo).

9. How big is the human embryo at Stage 23? _____ How old in days? _____

10. What has happened to the eyelids of the fetus at this stage? (It might be helpful to compare Stage 23 with Stage 22 to appreciate this.)

11. What is happening to the arms and legs at this stage? The website says that there is a "clear distinction of the subdivisions of the upper and lower limbs," but what does this mean? (Again, it is helpful to compare Stage 23 with Stage 22 here.)

ACTIVITY 3: THE VISIBLE EMBRYO

Now we'll take a look at some key changes during the 2nd and 3rd trimester of development. Go to Bio 21 on Canvas and under Lab 10, click on "The Visible Embryo". (Or go directly to www.visembryo.com/baby/index.html). What you see here is a spiral representing both fetal and maternal timelines during development. For this activity, we'll focus just on fetal development. **To get this information, you must click on each fetus image next to the spiral and not the spiral itself**.

12. On the chart below, indicate the timing (in weeks) of major changes to the fetus.

Key Development	Timing (in weeks)
Sex organs clearly visible	
Lungs produce surfactant	
Hearing is possible	
Bone marrow starts to make blood cells	
Fetus begins to develop its own Immune system	

13. What is surfactant? What important role dues it play in normal lung function?

ACTIVITY 4: DEVELOPMENT ON THE WEB

Using information from www.marchofdimes.com and Table 20.1 in your textbook, complete the following. (Hint: think about the stages most at risk with use.)

	Risk to fetus
Smoking	
Drinking	
Recreational Drugs	
Prescription Drugs	

ACTIVITY 5: STEM CELLS IN DEVELOPMENT AND ADULTHOOD (Dr. Katie Wilkinson)

Go to Bio 21 on Canvas and under Lab 10 click on "The Nature of Stem Cells" (or go directly to `learn.genetics.utah.edu/content/stemcells/scintro/`)

14. Which type of cells can give rise to any type of cell in the body and where are these found?

15. After about 1 week, the blastocyst is formed. Draw a blastocyst and label which cells will become the placenta and which cells will become the fetus.

16. After ~2 weeks, 3 layers of cells form. Name these 3 layers and what tissues they will form.

17. What causes stem cells to differentiate into different types of cells?

18. Where can you find stem cells in adults?

19. What is the function of adult stem cells (also called somatic stem cells)?

20. How do adult (somatic) stem cells differ from embryonic stem cells?

ACTIVITY 6: ETHICS OF OBTAINING STEM CELLS FOR RESEARCH (Dr. Katie Wilkinson)

There is an ongoing debate about the ethical use of stem cells in research and the clinic. In 2001, President Bush restricted federal funding for stem cell research to the existing embryonic stem cell lines (these restrictions were overturned in 2009 by President Obama). California voters approved Proposition 71 in 2004, which allocated $3 billion over 10 years to support stem cell research, including the creation of new embryonic stem cell lines. The California Institute for Regenerative Medicine (CIRM) was formed to disburse the money (www.cirm.ca.gov/). CIRM's mission is: "*To support and advance stem cell research and regenerative medicine under the highest ethical and medical standards for the discovery and development of cures, therapies, diagnostics and research technologies to relieve human suffering from chronic disease and injury.*"

Read the following three articles and answer the questions below. (Links for articles are also on the Bio 21 Canvas site under Lab 10).

"Stem Cell Quick Reference" learn.genetics.utah.edu/content/stemcells/quickref/

"The Stem Cell Debate: Is it Over?" learn.genetics.utah.edu/content/stemcells/scissues/

"Myths and Misconceptions about Stem Cell Research" www.cirm.ca.gov/our-progress/myths-and-misconceptions-about-stem-cell-research

21. What is the source of embryonic stem cells for research?

22. Are aborted fetuses a source of embryonic stem cells? Why or why not?

23. Fill out the following table with what you see as the pros and cons of using embryonic stem cells versus induced Pluripotent Stem Cells (iPSCs).

Stem Cell Type	Pros	Cons
Embryonic		
iPSC		

24. Do you agree with CIRM's decision to fund research into all types of stem cells, even embryonic stem cells? Why or why not?

ACTIVITY 7: STEM CELLS IN THE CLINIC (Dr. Katie Wilkinson)

Many researchers believe that stem cells hold great promise for treating a variety of diseases from spinal cord injuries to HIV/AIDS. To learn how stem cells can be used therapeutically go to `learn.genetics.utah.edu/content/stemcells/scfuture/` (also on the Bio 21 Canvas site under Lab 10).

25. What parts of your body can regenerate or grow again?

26. What is regenerative medicine?

27. What 2 types of stem cells are pluripotent and can differentiate into any cell in the body?

28. Name two strategies for using stem cells in regenerative medicine?

29. How are stem cells similar to cancer cells?

Now let's see some treatments being developed here in California using stem cells. Go to `www.cirm.ca.gov/patients/disease-information` (also linked on the Bio 21 Canvas site under Lab 10). Under **Disease Programs**, select a disease to learn more about using fact sheets and videos.

30. What disease did you choose to learn about?

31. How are researchers trying to use stem cells to treat the disease you have chosen? (Use back to page to complete your response.)

Human Biology: Lab 10

BIOLOGY 21
LAB 11: GENETIC VARIATION & MEIOSIS

Our biological legacy is contained in our genes. That is, we owe our traits and characters to our parents, grandparents, and so on. The study of genetics and inheritance provides fascinating insight into why we look and, perhaps, behave, the way we do. For example, we can use pedigree charts to trace certain traits back several generations. We can also use principles of inheritance to predict what our children will look like.

ACTIVITY 1: MEIOSIS ON THE WEB

Go to Bio 21 on Canvas and then to Lab 11. Click on "Meiosis on the Web" or go directly to `www.pbs.org/wgbh/nova/baby/divide.html` Answer the following:

1. What is the main difference between mitosis and meiosis?

2. What "problem" does meiosis solve? (Hint: why is meiosis called "reduction division"?)

Using information from this website, identify each step (the first one is listed for you) and describe what occurs at that step for mitosis and meiosis. Diagramming may be useful here.

	MITOSIS	MEIOSIS
Step 1 Name: Before We Split		
Step 2 Name:		
Step 3 Name:		

MORE>

Human Biology: Lab 11 11.1

	MITOSIS	MEIOSIS
Step 4 Name:		
Step 5 Name:		
Step 6 Name:		
Step 7 Name:		
Step 8 Name:		

	MITOSIS	MEIOSIS
Step 9 Name:	These steps are not found in mitosis	
Step 10 Name:		
Step 11 Name:		
Step 12 Name:		
Step 13 Name:		

MORE ▶

Human Biology: Lab 11

	MITOSIS	MEIOSIS
Step 14 Name:	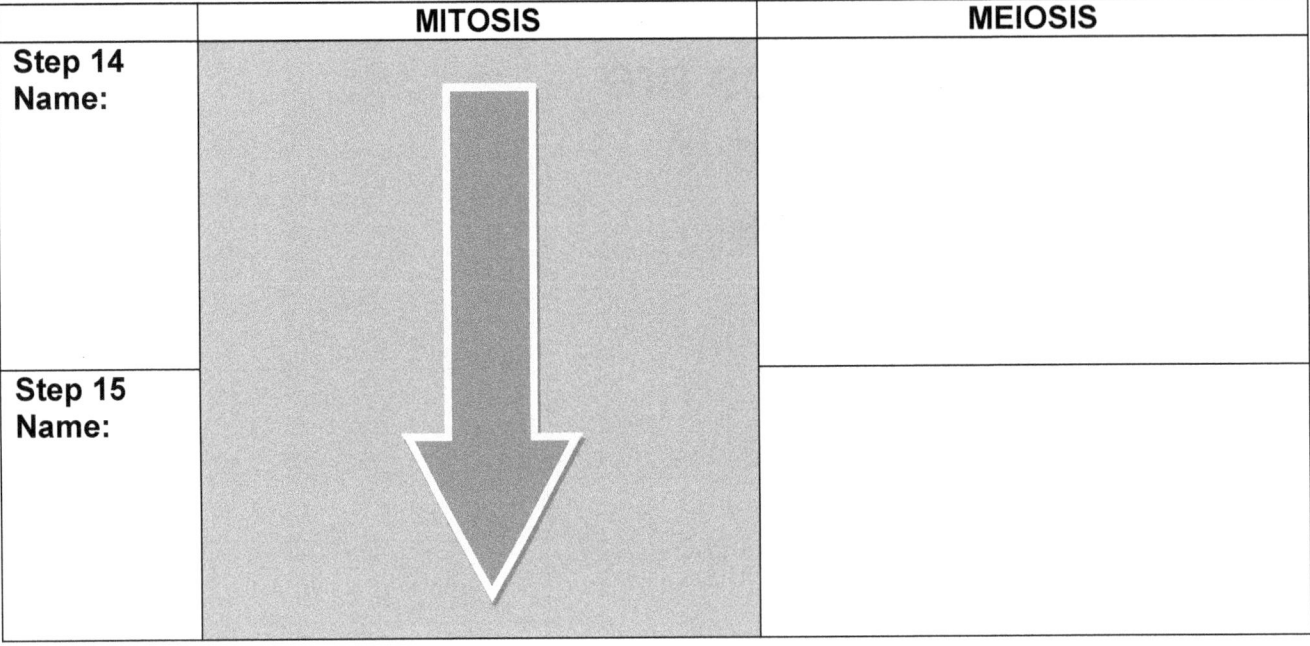	
Step 15 Name:		

ACTIVITY 2: TERMINOLOGY

The key to understanding inheritance and genetics is knowing the "language." Before you proceed with this lab, define the following terms (see Chapter 21 and the Glossary of your textbook):

Term	Definition
3. allele	
4. dominant (allele or trait)	
5. gene	
6. genotype	
7. heterozygous	
8. homozygous	
9. phenotype	
10. recessive (allele or trait)	

ACTIVITY 3: THE PUNNETT SQUARE EXPLAINED

Reginald C. Punnett developed a special chart to show the possible combinations of traits that can arise when two organisms are bred or crossed. The chart helps us to determine the probability of having a particular characteristic. The most basic Punnett Square contains four squares. One parent's gene combination is put across the top, while the other parent's is put down the side (see below).

In this example, we will find out how many black haired children a man with black hair and a woman with blonde hair could have. The gene combination, or **genotype**, for a person with black hair is **BB**. The genotype for a person with blonde hair is **bb**. This information is indicated in this Punnet square:

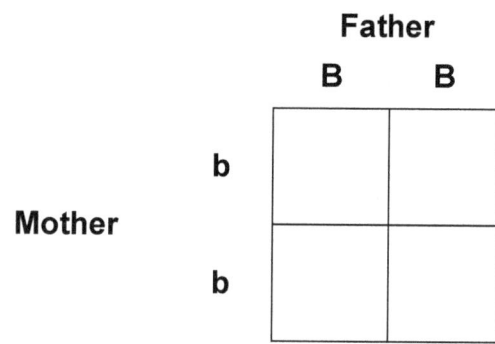

Next, carry the mother's alleles across the rows and the father's alleles down the columns.

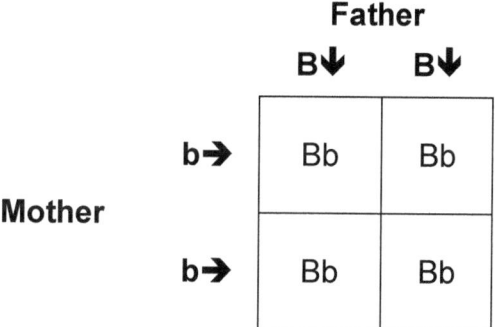

The big "**B**" is called the **dominant gene**, or the gene that will be expressed in the child. The little "**b**" is called the **recessive gene**, and will only be expressed if a dominant gene is not present. So, if a child inherits a "B" from his father, and a "b" from his mother, the "B" gene will be expressed and he will have black hair, like his father. In this case, the chances of having a child with black hair (**Bb**) are 4 out of 4 or 100%. It is important to point out that Punnett Squares give us only the probability of certain combinations occurring. As every gambler knows, the outcome sometimes can be wildly different!

STATION 1: PUNNETT SQUARE APPLIED

At this station, you'll find a list of characteristics and the genotypes that code them. Choose two of these characteristics and fill out the Punnett Squares below to predict what the offspring would look like. For example: If you choose the genotype for non-red hair (**Rr**) and the genotype for red hair (**rr**), you would plug these genotypes into a Punnett Square and determine the proportion of children (according to probability) that could have red hair and the proportion that could not have red hair. (The gene for red hair is recessive, by the way.)

Human Biology: Lab 11

Using red hair as an example where the father is **Rr** and the mother is **rr**:

Father

Mother

11. What proportion of children could have at least one copy of the dominant gene? _____

12. What proportion of children could have red hair? _____

13. What proportion of children could have two copies of the recessive gene? _____

14. What proportion of children could not have red hair? _____

Do the same for two more traits. For each trait, pick a genotype for you and one for your partner. Fill out the Punnett Squares below and answer the corresponding questions.

Trait 1: _____

Dominant allele: _____ *Recessive allele:* _____

Father

Mother

15. Proportion of children with the recessive allele? _____

16. Proportion of children with the dominant allele? _____

17. Proportion of children expressing the trait? _____

Trait 2: _____

Dominant allele: _____ *Recessive allele:* _____

Father

Mother

18. Proportion of children with the recessive allele? _____

19. Proportion of children with the dominant allele? _____

20. Proportion of children expressing the trait? _____

STATION 2: ARE YOU A TASTER?

About 70% of the population gets a bitter taste from a substance called phenylthiocarbamide (PTC). It is tasteless to the rest. The "taster" allele is dominant to the non-taster allele.

To determine if you are a taster, touch a piece of taste-test paper to the tip of your tongue. If you are a taster, you will detect it immediately. Otherwise, the paper will be tasteless.

21. Are you a taster? _____ What about your lab partner? _____

Please dispose of the PTC paper in the Biohazard container. If you want, help yourself to a piece of candy to get rid of the taste.

22. What are the possible outcomes of a mating a male who is heterozygous for the PTC tasting trait and a female who cannot taste PTC? (Use the letter **T** for taster, and **t** for non-taster. Then do the cross to answer these questions.)

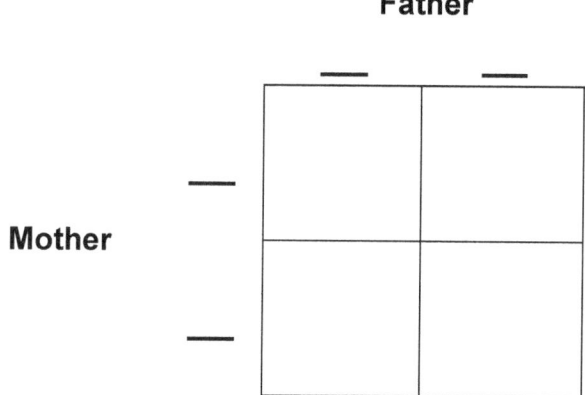

ACTIVITY 4: GENETIC DISORDERS

There are over 6000 known genetic disorders. Here we have you investigate two.

Fragile X Syndrome: www.fraxa.org (also on Bio 21 Canvas site under Lab 11).

23. What is Fragile X Syndrome? How prevalent is it (according to the website)?

24. What are some symptoms?

25. How is Fragile X inherited? Why can't a father pass Fragile X syndrome to his sons?

MORE▶

ACTIVITY 4: GENETIC DISORDERS (continued)

26. Why is Fragile X more common in boys?

27. When Fragile X is present in girls, why aren't girls usually as severely affected (you'll have to do some critical thinking here to figure this one out).

Klinefelter Syndrome: `men.webmd.com/tc/klinefelter-syndrome-topic-overview`

28. What is Klinefelter Syndrome? How prevalent is it (according to the website)?

29. What are some of the characteristics (or symptoms) of Klinefelter Syndrome?

30. What are some of the treatments for Klinefelter Syndrome?

ACTIVITY 5: GENETICS & INHERITANCE PROBLEMS

On the next several pages, you'll find many different genetics and inheritance problems. Yes, some of it may seem repetitive, but the more you practice these problems, the better you'll grasp this week's topic.

COMPLETE DOMINANCE

A completely dominant gene will totally mask its recessive allele.

Example: What type of offspring would be expected from the mating of a female homozygous for hanging ear lobes (EE) with a male with attached lobes (ee)?

Parental   Genotypes EE   ee

             Gametes   (E) (e)

F_1 Genotype   Ee

             Phenotype   hanging lobes

Note that the dominant gene is always written first. Since the offspring are all the same, there are no genotypic or phenotypic ratios.

Example: What type of offspring would be expected from the mating of a female heterozygous for hanging lobes with a male heterozygous for hanging lobes?

This problem may be solved in the form of a "Punnett Square," in which the gametes are placed in the form of a multiplication table.

Ova

Sperm	E	e
E	EE	Ee
e	Ee	ee

Phenotypic ratio: 3 hanging lobes to 1 attached lobe

Genotypic ratio: 1 EE to 2Ee to 1 ee

MORE>

Problems:

Note: empty Punnett Squares can be found on the last page of this lab. For most of these problems, you need to use the Punnett Squares in order to solve the problem. Fill them out first with the information given in each question, and then answer the question.

31. Some people have freckles on their face. Some freckled individuals, when mated to other freckled individuals, have only freckled children. Other freckled people, when mated to freckled persons, have some children with freckles, some without. When both parents are without freckles, no offspring have freckles. Explain the results. Write the genotypes for these crosses using the symbols **F** and **f**.

a. Genotype for no freckles: _____

b. Genotypes for freckles:_____

c. What kind of cross would produce only freckled children? _____

d. What kind of crosses would produce some freckled children and some children without

freckles? _____

_____

INCOMPLETE DOMINANCE

In some instances, the heterozygous form is phenotypically unique, displaying a condition intermediate between the two homozygous conditions. Since neither gene dominates in the heterozygous condition, this is referred to an incomplete dominance.

Example: Let **N** represent the allele allowing normal metabolism of the sugar galactose. Let **n** represent the gene for lack of ability to metabolize galactose due to the lack of an enzyme. The genotypes and phenotypes possible are as follows:
- **NN** normal
- **Nn** reduced ability to metabolize galactose (low level of enzyme)
- **nn** inability to metabolize galactose (lack of the enzyme; disease = galactosemia)

Problems:

32. A normal woman (**NN**) reproduces with a man with the reduced ability to metabolize galactose (**Nn**). What are the expected phenotypic and genotypic ratios of their offspring? What is expected when two persons with reduced metabolic ability mate?

a. Genotypic ratios:_____

b. Phenotypic ratios: _____

c. Expected outcome when two people with reduced metabolic ability mate _____

MORE ▶

33. Glycogen-storage disease (excessive accumulation of glycogen in the kidneys and liver) results from the homozygous condition **gg**. **Gg** results in abnormally high level of some sugars in the blood. **GG** is normal. A woman with high sugar levels, and whose father had the full disease, reproduces with a man with high sugar levels, although his mother was normal. What are the genotypes of these people? What are the possible genotypes of the woman's mother and the man's father? Could the woman have any brothers with the disease, or any who are normal? Could the man have any sisters with the disease, or any who are normal? What are the expected phenotypic and genotypic ratios of the younger couple's offspring?

a. Genotype of the woman_____   b. Genotype of the man _____

c. Genotype of woman's father_____   d. Genotype of man's mother_____

e. Possible genotypes of the woman's mother _____

f. Possible genotypes of the man's father _____

g. Could the woman have any brothers with the disease, or any who are normal? _____

h. Could the man have any sisters with the disease, or any who are normal? _____

i. Expected genotypic ratios of the younger couple's offspring: _____

j. Expected phenotypic ratios of the younger couple's offspring: _____

PROBABILITY
It must be emphasized that in these exercises you are determining the expected results – the proportions of different types of offspring which would be formed provided they follow idealized laws of probability. The actual results may be different. The following exercises will demonstrate this.

34. If you toss a coin in the air ten times, you would expect it to come up heads five times and tails five times. Do this and see what you get.

_____ heads   _____ tails

35. Take two coins and let heads represent the dominant allele "E" for freely hanging lobes, and tails represent its recessive allele "e" for attached lobes. One coin will represent a heterozygous father, the other coin a heterozygous mother. The sides which appear will indicate the allele in the sperm and the allele in the ovum which unite in fertilization.

Of four children, you expect:

a genotypic ratio of _____ EE: _____ Ee: _____ ee

a phenotypic ratio of _____ hanging lobes: _____ attached

Toss the coins simultaneously four times and record your results, noting the genotypes and phenotypes of the offspring (paired coins) after each toss.

MORE ▶

Human Biology: Lab 11   11.11

Actual results

_____ EE: _____ Ee: _____ ee

_____ hanging lobes: _____ attached

SEX DETERMINATION

The primary determinants of the sex of an individual are specialized sex chromosomes. In humans, these sex chromosomes are of two types: X and Y. The X chromosome promotes femaleness, and the Y chromosome promotes maleness. An individual with two X chromosomes (XX) is a female and an individual with one X and one Y (XY) is a male. The Y chromosome possesses very few known genes. In contrast, the X chromosome has over 800 genes, most of which do not have homologous alleles on the Y chromosome (which again, has very few genes). Because of this, the genes on the X chromosome are said to be "sex linked".

	Female	Male
Parents	XX	XY
Gametes	X X	X Y

Offspring

	X	X
X	XX	XX
Y	XY	XY

36. Which parent determines the sex of the offspring? _____

37. If a gene is on the X chromosome of the father, what percent of his daughters will inherit the gene? _____

38. What percent of his sons? _____

39. What percent of the sons receive a Y chromosome from the father? _____

40. What percent of the daughters receive a Y chromosome from the father? _____

MORE▶

SEX-LINKED FACTORS

Any gene carried on an X chromosome is said to be sex-linked. Therefore, females have three possible genotypes, while males have only two since we may consider the Y chromosome as carrying no genes.

Example: Let **N** represent the gene for normal vision. Let **n** represent the gene for red-green color blindness. These genes are sex linked. The possible combinations are as follows:

Female
XX

POSSIBLE GENOTYPES	PHENOTYPE
NN	Normal
Nn	Normal (but a carrier)
nn	Color-blind

Male
XY

POSSIBLE GENOTYPES	PHENOTYPE
NY	Normal
nY	Color-blind

Note that the Y chromosome doesn't have alleles, so we just use "Y" as a place holder.

What would be the expected types of offspring from the mating of a homozygous normal woman with a color-blind man?

Female

		N	N
Male	n	Nn	Nn
	Y	NY	NY

All offspring would have normal vision. However, the females would be carriers for color-blindness.

Problems:

41. The gene for the ability to smell cyanide (**C**) is dominant over the gene for the lack of such smelling ability (**c**). The genes are sex-linked. A woman who cannot smell cyanide mates with a man who can.

Will their sons be able to work as safety inspectors in the cyanide factory? _____

Will their daughters be able to work as safety inspectors in the cyanide factory? _____

42. Hemophilia is a malfunction in which blood fails to clot. The gene producing this (**h**) is recessive to the gene for normal clotting (**H**). The genes are sex linked. If a normal woman mates with a normal man, and they have a hemophilic son, which parent is responsible for the malady?

MORE➤

Human Biology: Lab 11

a. Responsible parent_____

If this son later reproduces with a normal female whose father was hemophilic, what type of offspring might be expected?

b. Expected offspring of hemophilic son and normal female with hemophilic father (hint: since this is a sex-linked trait, you have to state the outcome for girls and the outcome for boys):

_____

_____

PUNNETT SQUARES (FOR PRACTICE)

CPSIA information can be obtained
at www.ICGtesting.com
Printed in the USA
LVHW100113290819
629362LV00006B/107/P